Masoud Asari

Principles of Village Design in Architecture

Masoud Asari

Principles of Village Design in Architecture

with a Sustainable Energy Perspective

Noor Publishing

Imprint
Any brand names and product names mentioned in this book are subject to trademark, brand or patent protection and are trademarks or registered trademarks of their respective holders. The use of brand names, product names, common names, trade names, product descriptions etc. even without a particular marking in this work is in no way to be construed to mean that such names may be regarded as unrestricted in respect of trademark and brand protection legislation and could thus be used by anyone.

Cover image: www.ingimage.com

Publisher:
Noor Publishing
is a trademark of
Dodo Books Indian Ocean Ltd. and OmniScriptum S.R.L publishing group

120 High Road, East Finchley, London, N2 9ED, United Kingdom
Str. Armeneasca 28/1, office 1, Chisinau MD-2012, Republic of Moldova, Europe
Printed at: see last page
ISBN: 978-620-5-63470-7

Principles of Village Design in Architecture with a Sustainable Energy Perspective

By

Masoud Asari

Master of Architect. Bam Branch, Islamic Azad University, Bam, Iran

Dedicated to the merciful angels who:

The lord of the worlds, who began to guide his servants with the teaching of the pen.

My parents, whose presence is a crown of honor for me and their name is a reason for my existence because these two existences after the lord, have been the source of my existence, took my hand and taught me to walk in this valley full of ups and downs.

Content

Chapter I

Introduction

Introduction

A look at the stages of historical development of many cities in our country leads us to the fact that although urban development in them was accompanied by quantitative development of the city and an increase in construction along with an increase in urban incomes, but in terms of environmental quality, this development has been associated with many challenges. One of the constant problems of our cities today is the lack of recreational and tourism spaces.

Tourism has become the basis for idea generation, entertainment and recreation, job creation and entrepreneurship, and in a more modern form has become an interconnected network of economic, social, civil and cultural infrastructures. Meanwhile, the vast country of Iran, with its extensive cultural, heritage and natural attractions, is one of the top countries in the world in terms of tourist attractions and has many potentials for the development of tourism, tourism and leisure at the national level and is international.

Figure 1. Village

Mazandaran province with many pristine and natural areas and many God-given blessings and unique climate in the country, and proximity to the Caspian Sea and its Alborz slopes, including the most innovative and fun tourist areas in the country and even Iran is a neighboring country and the annual return of millions of travelers to this province indicates the high position of this province in the country's tourism system and the amount of attention paid by its compatriots to this province and its capabilities. The site of this project, located in Amol city and also located near Haraz river, all the capabilities and potentials of the green region of Mazandaran, including the benefit of favorable climate, greenery and vegetation density, proximity to the Caspian Sea, welcome travelers and ...

Generalities

Problem statement

Appropriate response to the physical, mental and spiritual needs of human beings is one of the necessities that public spaces should be organized in the form of various and specific functions in this direction. Lack or weakness of social functions in this regard leads to inefficiency in human beings and its negative consequences will affect the community. The growth and development of urban life and the increase in demand for housing and buildings have caused the environmental quality value to be ignored and the destruction of natural and environmental spaces, especially urban green spaces, and in practice, urban spaces become more financial and economic interactions.

Human beings today, understanding the importance and position of public spaces and areas of recreation and recreation in daily life and the various functions of these spaces among the structure of human gatherings, strive to preserve and revive and make good use of available resources in the planning agenda and Has set its own policies. On the other hand, the lack of attention to socio-cultural approach in the design of such spaces has caused their developmental role to take on an individual aspect and weaken social interactions.

Today, the big cities of our country are far from having the necessary limits and dimensions in having social spaces, especially in the field of recreational spaces. Hence, the lack and absence of these spaces will create an unbalanced spirit and an unbalanced pattern of behavior in citizens. Another prominent aspect of the development of recreational and leisure spaces is its economic effects on the exploiting and exploiting society. Impacts on job creation as well as national income through the creation of national and international recreation and tourism areas. Today, it has been considered by most managers and decision-makers in the world, and therefore, in recent decades, we have faced significant growth and development of tourism spaces in most countries.

In this regard, the managers and officials of the holy system of the Islamic Republic of Iran, in order to develop the culture of tourism and also to create suitable tourism and recreation areas in the country, have provided appropriate arrangements and contexts and in recent years, support for this sector significantly has increased. This, in addition to the positive feedback that will bring in the psycho-social spheres, will create the necessary conditions for the development of the tourism industry and subsequently earn national income through the development of the tourism industry, especially in the national and international dimension. The necessity of this arises from the fact that at present, there is a significant shortage in the field of large-scale national and international promenades in the country and the need to develop this sector is undeniable and necessary.

Figure 2. Village

On the other hand, the creation of recreational and leisure spaces in the inner city, due to the lack of suitable pristine and empty spaces in the inner city, as well as problems such as traffic and pollution caused by the creation of inner-city travel, today is facing many problems. The creation of tourism and recreational spaces in the suburbs of cities and in pleasant climates located in the foothills of nature, is more welcomed. What we are witnessing in recent years is the great popularity of the residents of the big cities of the country, for suburban trips and turning to recreational-recreational spaces outside the cities. However, recreational-leisure spaces within the city are not so welcomed, because the citizens want to leave the city more than anything.

A noteworthy point in this regard is the climatic diversity and richness of the natural environment of the expansive country of Iran. This capability has created a lot of potentials for the development of ecotourism and nature tourism for our country, which, considering the wide popularity of the people, can provide a comprehensive development infrastructure that, if properly managed resources and capital, its positive feedback. It will be shared by the Iranian society. In this regard, green spaces and recreational-natural recreational spaces outside the city, which are located in the field of human-nature interactions as well as human-human, outside the structural framework of the financial-economic system of large cities, are more important. Among these very important areas in our country, we can mention the green area of Amol city that the site of this project, with its location in the middle of this area, is one of the points with similar potentials.

Meanwhile, macro-national policies, based on national planning plans, have placed the northern lands of the country and the provinces of Gilan and Mazandaran to play recreational and ecotourism roles. The region should move towards approved macro policies and functions that play recreational-ecotourism and ecotourism roles.

Meanwhile, the strategic position of Mazandaran province and its high potentials in attracting tourists, has caused that this province, despite the low level of investment and lack of tourism development programs in the province, annually receives numerous clients and tourists from different parts of the country and Even other countries. Meanwhile, the site of this project is no exception to this rule and it is obvious that if properly planned and designed, it will certainly strengthen the recreational-ecotourism and ecotourism role defined for the province, taking advantage of the existing advantages. It was able to play a role on a national, regional and international scale.

Figure 3. Stream northern lands

Establishment of various functions and uses in the project site should be done according to the characteristics of the natural and artificial environment as well as its socio-cultural and economic feedback and the occurrence of social and environmental anomalies in the area should be avoided. This has revealed the need for more comprehensive and practical attention to the future development of the project site, and from this perspective, the preparation of development plans with functional, social, environmental and economic considerations will be part of the project plans so that the collection can be considered and Planned to meet the needs of the region.

Importance of the research topic

The tourism industry has long been with the aim of visiting historical and natural attractions (ecotourism), pilgrimage, trade, use of medical services and . . . It has been formed and gradually developed with the evolution of equipment and the provision of more suitable travel facilities. Today, this industry is considered as one of the most lucrative activities by all countries. In recent times, due to the development of urbanization and the growth of industry, environmental pollution has increased and has caused the demand for travel and leisure in urban communities to increase.

To meet the demand for day-to-day travel, we are witnessing the growth and development of information and communication technologies and services, services, facilities and equipment in the tourism complex in the world. Because in most countries that have long-term plans to attract tourists, tourist villages have special standards. These studies are necessary for better planning and prevention of many problems in the development of the tourism industry. By studying the tourist villages and also the conditions of their formation, the necessary criteria can be examined and this will lead to the correct design of the tourist village. In this research, by studying foreign cases and comparing them with similar domestic examples, we have achieved the necessary criteria for designing a recreational complex in tourist villages and by applying it in a practical case study, we will address the necessity of this research.

Figure 4. How would you design an urban eco village?

The ultimate goal of this project is to design and improve the quality of the tourist village by creating recreational and cultural spaces. Including an efficient fashion system, with appropriate economic return for society, in order to play a recreational-leisure role, with environmental, functional and socio-cultural considerations, in order to spend leisure time and to improve the mental, physical and mental dimensions of the population. From it and improving the environmental quality of the project area, in

Amol city located in Mazandaran province. From this perspective, the following operational objectives can be proposed in relation to the issue:

Environmental goals

The macro-environmental goal of this project is to provide a physical plan and design system appropriate to the capabilities and environmental capabilities of the project in order to preserve and protect and prevent the destruction of natural lands in the project area and promote its role and environmental position in the limited environmental life. It has a direct impact that can be pursued through the following operational objectives

- ➢ Reducing the destructive role of development interventions at the land level located within the project area.
- ➢ Creating the necessary bases for the protection and sustainable preservation of natural resources by the people.
- ➢ Creating the necessary conditions for the presence of managed people in national and natural lands in order to prevent the process of destruction of environmental values in the region.
- ➢ Development of green spaces, including afforestation, tree planting, etc., as well as water spaces, including waterways, ponds and reservoirs, in order to soften the air and improve the environmental indicators of the project bed.

Functional goals

The main objective of this project is to provide an optimal physical performance plan for the project lands in relation to the capabilities and limitations of the region in order to create a tourist village to meet the needs and shortcomings in the recreational and ecotourism areas. The following operational objectives will be achieved.

Figure 5. Tourist village

- ➢ Analyzing the current functional status of the site and its communications and interactions with neighboring functions and the region index.
- ➢ Improving the qualitative and quantitative level of site performance in relation to its capabilities and limitations and in relation to and interaction with other spaces and recreational, leisure and sports functions in the province.
- ➢ Creating suitable and worthy landscapes and mirrors on the banks of Haraz river.
- ➢ Establishment of a new center for recreational and leisure activities in Mazandaran province, due to the existing shortcomings.
- ➢ Responding to recreational-leisure needs in the province, by establishing the required functions at an appropriate scale.

Social, cultural and economic goals

The social, cultural and economic purpose of this design is an optimal social, cultural and economic system appropriate to the project design site and site in relation to the demographic structure and social and economic position of the users of the complex and reducing anomalies and promoting social norms at domestic and foreign levels. It is related to the project site and also creates new economic and job opportunities, in a way that includes positive economic feedback for the project community. This will be pursued through the following operational objectives:

- Psychological refinement of the users of the project area of influence by creating a happy and fun atmosphere based on a suitable design.
- Creating new job and economic opportunities in order to improve the economic potential of the region's residents by attracting tourists along with creating jobs for the people of the region and designing local markets.
- Responding to the recreational-leisure needs of clients in order to regain physical, mental and spiritual strength by designing fun spaces.
- Improving the qualities and indicators of control security and reducing anomalies and social crimes within the site.
- Creating the necessary social and cultural backgrounds to prevent the destruction of the environment and pristine and natural space of the project area and the importance of crossing the river through the heart of the city by highlighting the importance and vital role of water in human life by combining water and architecture.

To achieve the main goal (design of a tourist village) according to the sub-objectives (physical program, climate studies, study of effective parameters, etc.) steps must be taken.

In such a way that it can play a recreational-leisure role with environmental, functional and aesthetic considerations in order to spend leisure time and with the aim of improving the mental, physical and mental dimensions of the population using it and also preserving pristine and natural spaces.

Figure 6. Village

Research background or related literature

The history of Iranian campuses should be traced back to ancient times. At the same time, the garden of cities that was created in Iran during the Safavid period, and especially in Isfahan, inspired "Ebenezer Howard", the founder of the theory of city gardens. Obviously, the construction of parks and gardens in the style of European countries in Iran began in the 1971. (Ebrahimzadeh, 2: 2008)

The beginning of attention to urban green space in the West in the meaning and concept of the new industrial age should be sought in the era of the industrial revolution and the resulting changes in the political, economic, social and cultural dimensions. (Ebrahimzadeh, 3: 2008)

Comfort and creating the necessary conditions for a desirable life has always been the goal of human beings. Changes have been made on the planet by human beings in order to create human comfort and well-being. Initially, urban life began with the use of traditional methods to continue living alongside nature, but with the industrialization of urban communities significantly expanded and transformed. Became natural energy mines and forests. We must now seek to create the conditions for harmony with the environment and the creation of a sustainable environment. (Rahshahr, 2003: 6)

In this dissertation, after stating the generalities, considering the importance of the subject to answer the questions, considering the objectives in the second chapter, we will express the generalities and theoretical foundations of the tourist village in order to gain a more comprehensive knowledge about it. We look at the existing and the values and potentials of the region to reach a deeper thought for design, and the fourth chapter is the presentation of the physical program. For accurate and principled design, and finally the fifth chapter examines the climatic conditions and analysis of the site in order to help a proper form and correct placement on the site with the help of these insights.

Chapter II

Generalities and Theoretical Foundations of the Plan

Introduction

In this chapter, after describing some of the common theories and models in relation to the tourist village, we come to a comprehensive knowledge of the meanings and concepts, principles and related theoretical theories that actually depict the semantic mentality of the project and at the same time, after examining the case examples at the general and micro levels of the spaces, we arrive at the necessary solutions and suggestions for assessing the needs of the spaces and the form of communication between them, which we finally use in the design process.

Definitions of concepts and theories related to the subject

First, we will define the words related to the title, such as entertainment, recreation, promenade, excursion, in order to reach a general acquaintance in this regard, and then we will describe the related titles and their application in the dissertation design section.

Fun

Recreation is any activity or inactivity that is done with the previous intention and with desire in leisure time. Therefore, recreation is an emotional and enjoyable experience that is given to people by their desire in their free time. In other words, when there is not the slightest feeling of compulsion in activity or inactivity in leisure time, this state can definitely be called entertainment. Entertainment is generally complete with cardboard, and its distinguishing feature is not its activity aspect. Rather, it is a state, a state and a feeling in which he touches people and learns them. Every pastime is done in his spare time, but not all of his leisure time is spent just for fun. Activity or inactivity that is without purpose and plan and without previous intention, is not considered part of entertainment. (Majnonian, 1995-16)

A researcher named Newmin offers this definition ("Recreation refers to any individual or group that takes place in leisure time and because it is a free, voluntary and enjoyable activity, has a special attraction") Carlsen, Deep and McLean in his book recreation in American life offers the following definition: "Recreation is any kind of enjoyable leisure activity in which people voluntarily participate and make themselves happy and entertained".

Recreation

Recreation means the evaluation of leisure time in order to reconstruct and self-discover the mental and physical self-recreation. In general, it means self-reconstruction (reconstruction and reconstruction) of the soul and body of a person to start a new period of activity-work. (Shafiee Nasab, 2004: 13)

Recreation is generally referred to as any kind of entertainment and game activity that is done in this field, whether indoors or outdoors. But all the games and entertainments that are usually done outdoors and outdoors, especially in natural resources, such as walking in nature, watching the scenery of forests and pastures, horseback riding, boating, skiing, camping, picnics and ... It is called recreation. (Tavakoli, 1992: 92)

Outdoor recreation requires space and resources. The most suitable resources that can improve the quality of recreation are natural and less modified resources that still retain their aesthetic aspects. Using all forms of parks, regardless of quality, extent or distance and accessibility, means recreation. (Majnoonian, 13: 1995)

Resort

A natural resource for recreation is any natural system, whether water or land, set aside for recreational use. A promenade can cover a wide range of natural features, from a simple stream to a deep cave. The use or possibility of using natural systems is a prerequisite for turning it into an actual or potential recreational resource. (Majnoonian, 13: 1995)

Rounds

Practical tour is a recreational activity that is informally related to consumption and motivation, time and place conditions, economic power, intellectual and cultural level, availability of facilities and personal interests about each individual or group (Haqqani, 4.: 2002)

Green space and its importance

Aspects of recreation and leisure today, green space has become a valuable national asset that serves the government as a source of capital and directly or indirectly, so that a very close and inseparable relationship between reducing medical expenses, increasing the level. There is public health, improving the quality of life and increasing the average life expectancy with the development of green space and tree planting. (Khademi, 6: 1986)

Figure 7. Notable large urban nature

According to the contents expressed in the design, the green space also plays an important role, including its importance in the intended use in the design, we can name the following items that we must address in the design:

A- The importance of green space in the fight against pollution.

B- The importance of green space in preventing noise pollution.

C- Temperature adjustment.

D- The effect of green space on people's morale.

E- Creative beauty.

C- Permeability and prevention of erosion.

Park

Parks are green spaces designed with different uses for public use, in terms of research, education, recreation and maintaining the health of the environment and people. In fact, parts and tissues of the city that the public has physical and visual access to and activities take place. In fact, it is a place to spend leisure time, interaction, conversation, education and so on. (Country Management and Planning Organization, 2001: 26) In his Ecological Glossary, Richard Carpenter defines the word "parkland" as "an area in which trees are scattered in groups or individually in a bed of grass cover."

This term is used today more than any other term to refer to natural lands with scattered trees and quasi-forests that have the necessary potential for a specific type of recreational activities. The following definitions in different cultures show a corner of the characteristics of the park.

Parks are usually spaces that have a variety of uses at the same time limited and indeterminate and are built with a flexible form and minimal construction and maximum use of natural materials for walking, relaxing, recreation, thinking, playing and .... (Izadpanah Jahromi, 11: 2002)

Figure 8. Park

According to the definition of the park, the existence of such a space is necessary in the design of the village, and parks with special titles such as children's parks, women's parks can be designed in different parts of it.

Water and architecture

Throughout history, our architects have incorporated water into their collections and used a treasure trove of physical, religious, and mythological features to enrich their architecture. Before the advent of Islam in Iran, architecture was present along the water and in the lap of nature without disturbing it. The role of water was more of an abstract role, shrines, temples of Anahita, fire temples, etc. were formed next to water and finally respect for the existence of water. It was as if water and fire coexisted peacefully and celebrated their grandeur. Examples of this can be seen in Azargashnast fire temple and Karian fire temple. Water plays a major role in the formation of the Iranian garden and trees, plants and flowers have the most important role after water, the presence of which has its roots in water. The presence of water in the Iranian garden shapes the character of the environment. (Imran Shahran Gilan, 2008: 49)

Ever since humans began building cities along the river, the composition and flow of rivers has led to the formation of street designs, alleys and parks. Parallel and vertical networking with the river axis spreads throughout the city and streets and blocks form a building. In fact, bridges act as intermediaries. In Isfahan, the Armenian region of Jolfa is located on the other side of the Zayandeh River and connects thirty-three bridges in this region to other parts of the city.

The site to be designed in Amol city is located in a place that is adjacent to the bridge in two directions and the river in one direction. Bridges often have a prominent urban location and if they are attractive, they will become a symbol of the city, such as Khajoo Bridge and 33 bridges and suspension bridges in Amol. Unlike fountains, springs and rivers, which are inherently kinetic elements, reservoirs and ponds are calm elements that collect water from the water cycle. Five bridges cross the Haraz River inside the city of Amol, all of which are bridges connecting the two main parts of the city.

As mentioned, they have an outstanding urban location, especially the suspension bridge next to the site for design, which can be used in the design of the arch design, such as bridge piers, and because of the historical values of bridges, it is better to design buildings in terms of match the height and appearance with them. Stagnant water naturally reflects images and, due to its reflectivity, is a determining factor in composition.

Figure 9. Water and architecture

Their mirror surface accepts the surroundings and then reflects. In our architecture, ponds are used as a symbol of stagnant water and are balanced with their regular geometric shapes. The ponds in front of the building complement the architecture and reflect them like a mirror, a clear example of which is Isfahan Chehelston. Another suggestion that can be expressed for the design of the area is the use of these ponds and fountains, which we combine with architecture by creating a failure in the river path and introducing water into the site.

Theories

Tourism (tourism and its types)

Tourism (WTO) as defined by the World Tourism Organization (WTO) refers to activities that go to places outside of their normal environment for leisure, work, and other purposes. Tourism in its broadest definition includes people who travel in connection with their work and profession and those who carry out scientific and research activities.

In this way, the scope of tourism's impact on the environment and its impact on the environment becomes much wider. Tourism is also a set of interactions that take place in the process of attracting and hosting tourists between travel agencies, governments of origin, host governments and local people.

Tourism is a multidimensional category and is related to several factors. Major and influential factors in the tourism industry are: tourists, countries of origin, destination governments, indigenous peoples, tourism organizations (domestic and international travel agencies), educational institutions (universities and technical and professional organizations), service organizations (hotels), hotels economic infrastructure (roads, sewerage network, communication networks), transportation network (air, land, sea), tourist attractions (historical, natural, cultural ...) all these factors under the influence of environmental factors such as cultural and social factors, Political and security, economic, technology and environmental. (Zahedi, 4: 2006)

Typology of Tourism (Tourism)

In this section, several types of tourism classification are mentioned so that after getting acquainted with its different types, it is possible to determine the type of tourism desired in this design and design it based on their needs:

Mass tourism

Mass tourism is a common commercial tourism that exists on a large scale around the world. Such tourism activities are carried out by large tourism companies and have their own markets, which are usually concentrated in certain areas of the world. Mass tourism varies in terms of travel motivation and destination of choice. In this type of tourism, tourists are looking for originality, novelty and diversity in the neighborhoods visited.

Figure 10. Tourism

Fam tourism or cultural tourism, ecotourism and adventure

The word fam consists of the initials of the words cultural, ecotourism and adventure. In this classification, tourism is divided into three categories. In cultural tourism, the tourist's goal is to visit the ceremonies, characteristics, and cultural manifestations of the host community. In ecotourism, visiting nature is in the center of tourists' attention. Adventure tourism is also associated with risk-taking and adventure.

Normal tourism and nature tourism

In this classification, tourism is divided into two categories: normal and naturalistic tourism. Although all types of tourism are related to nature in some way, but in this classification, the end use is not considered. Of course, drawing a line between different types of tourism is not possible and in some cases there is a possibility of overlap and overlap. For example, sports tourism overlaps with ecotourism and cultural tourism with historical tourism and pilgrimage tourism.

Normal tourism

Ordinary tourism is a type of tourism whose chosen destination does not necessarily have to do with nature. This type of tourism can in turn be divided into several sub-types.

A. Historical and ancient tourism

In this type of tourism, tourists pay attention to historical attractions and antiquities, such as: Persepolis in Iran, the Three Pyramids in Egypt, historical monuments in Rome and Greece.

B. Pilgrimage tourism

In this type of tourism, places of pilgrimage and religious ceremonies and rituals are considered by tourists, such as pilgrimage to the Kaaba, Jerusalem and Mashhad. Examples of religious ceremonies and rituals that attract the attention of some tourists are: Ashura mourning and carpet weaving.

C. Cultural tourism

Many tourists are interested in visiting and attending cultural and artistic ceremonies, and they adjust their itinerary in such a way as to be in harmony with the season of these rituals and ceremonies. Participation in art programs such as music, opera and theater concerts can also be included in this category.

Figure 12. Cultural tourism

D. Sports tourism

Some tourists, in order to get rid of the monotonous life throughout their work and occupation, are looking for a time and place to be required to do sports exercises. Sports tourism has a long history. The World Olympic Games are a clear example of sports tourism, which attracts a large number of tourists and provides a rich income for the host countries.

E. Meet relatives and friends

The purpose of some trips is to visit friends and acquaintances. Of course, in addition to these visits, there may also be visits to tourist attractions, but the main focus of these trips is to meet relatives and acquaintances.

Medical tourism

Natural tourism includes those tourism activities that deal directly with natural resources and attractions.

Natural tourism can be classified into six sub-categories:

A. Coastal tourism

B. Adventure tourism

C. Consumer tourism

D. Closed Tourism

E. Health Tourism

F. Ecotourism

Here is a brief description of each of these types:

A. Coastal or sand tourism

This type of tourism is closely related to the recreation of the Mediterranean coast, parts of the Pacific Ocean and Southeast Asia. In fact, this type of tourism with its uncontrolled development has caused negative economic, cultural, social and environmental effects in these areas. In Iran, the indiscriminate and reckless use of natural resources of the country's coasts and the imposition of contaminated waste and waste to these areas has caused irreparable damage to the environment of the region.

Figure 13. Coastal or sand tourism

B. Adventurous tourism

Adventure tourism is a type of tourism that is somewhat risky and requires physical activity and special skills. Examples of this tourism are: water skiing, mountaineering, sailing, caving, which in all of these activities, medicine, it brings effort and challenge, and these efforts are accompanied by excitement.

C. Consumer tourism

In all types of tourism, there is a general consumption factor. But consumer tourism refers to activities that lead to the consumption and damage of natural resources. The most common types of tourism are fishing and hunting. This type of tourism has faced many criticisms. Among other things, killing animals from an environmental point of view disrupts the biological balance.

D. Enclosed tourism

Enclosed tourism refers to the type of tourism that keeps elements of the natural environment under special and controlled conditions and exposes them to tourists, such as zoos, aquariums, and bird cages.

E. Health Tourism

Health tourism is a type of tourism that is related to health and medical activities and is considered by tourists who pay attention to the healing properties of natural resources such as: hot springs of medicinal plants, sludge beaches and medicinal grasses.

F. Ecotourism

There are several definitions of ecotourism, some of which are mentioned here:

1- Visit an area to see the land, animals and plants intact in that area.

2- Travel to nature in a way that while protecting the ecosystem, the dignity of local communities is also respected. In this definition, in addition to natural resources, the values of local people should be respected. In this definition, in addition to natural resources, the values of local people have been considered and the need to create a balance between natural resources, tourism, local community and tourists has been considered.

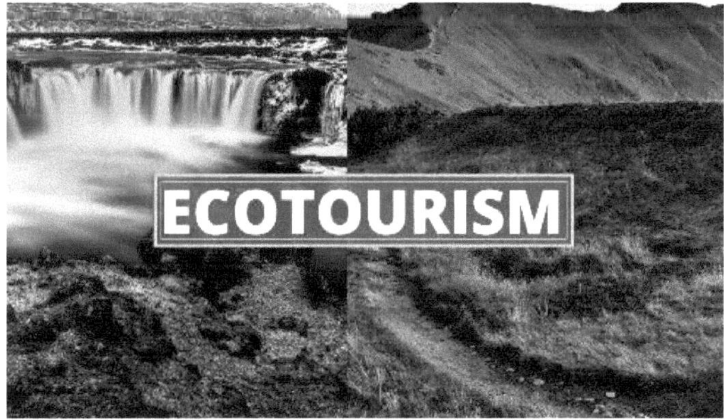

Figure 14. Ecotourism

3- Hector Ceballos-Lascurian Hector Ceballos-Lascurian was one of the first to define ecotourism in the 1980s. "Ecotourism means traveling to almost untouched natural areas with the aim of learning, admiring and using natural landscapes, wildlife, as well as the cultural past and present of indigenous peoples," he said.

4- The Canadian Environmental Advisory Council recognizes ecotourism as a nature travel experience that helps protect the ecosystem. At the same time, it preserves the host community.

5-The definition provided by the Australian National Ecotourism Institute is: "Natural tourism, which combines education and environmental awareness and is managed in an environmentally sound manner."

6- The International Union for Conservation of Nature (IUCN) defines ecotourism as follows:

"Ecotourism is the responsible travel to relatively pristine natural areas in order to enjoy nature in a way that has few negative effects on nature and provides the basis for socio-economic participation of the indigenous population."

7- The International Ecotourism Society (IES) definition of ecotourism is:

"Ecotourism is responsible travel to natural areas that protects the environment and ensures the well-being of local people." (Zahedi, 2006: 4-17) According to the explanations given, the type of tourism that should be considered for the design of the desired village is naturalistic tourism of the ecotourism type and part of the coast, considering that environmental protection is one of our design approaches.

Generalities and definitions of tourist village

In order to conduct studies on the design of a tourist village, we must first examine the generalities and concepts of tourist villages, their types and functions, and the conditions for their creation. In this part of this research, we examine the concepts of tourist villages with reference to foreign examples.

Definition of tourist village

There are places in a province that place or distances close to it in terms of natural, cultural and historical attractions have the necessary capabilities to attract tourists and need to establish facilities and facilities for tourism and information.

The tourist village, as a center for the concentration of domestic tourists in order to make leisure, holiday and residential trips within the province to various tourist attractions throughout the province, can be a center for accommodation and distribution of tourists in each province. Therefore, this center should have all the features needed by residents, tourists and those who spend their leisure time in the village. Therefore, the construction of recreational, sports, service and welfare spaces in line with world standards today is envisaged to create a village. Environmental considerations and sustainable development are other features that should be considered in the proposed plans.

Figure 15. Tourism villages

In this case, we can be sure that the construction of such villages will have a good return, both economically and culturally. The goals that the tourist village is supposed to be known for are:

1- Increasing the number of tourists in an area.

2- Increasing public revenue from tourism development.

3- Increasing job opportunities.

4- Realizing the potential tourism capacities in a region.

5- Attracting more private sector participation in tourism development.

6- Increasing investment attraction.

7- Improving the level of public services to tourists (Tourism Zones Organization, 2002).

Tourist villages are areas that have some characteristics of a place as a subject of tourism. In this area, the customs and culture of the local communities are still untouched and pristine. The tourist village is also a combination of some protection and conservation factors in which, in addition to preserving nature and the environment, items such as cooking methods, traditional agricultural systems and local communities are preserved. Creating and providing suitable facilities for turning a place as a tourist destination is of special importance. These places have spaces or spaces have been created in them that make visitors experience interesting and memorable moments in a space different from where they live. In the following, we will introduce the types of tourist villages to determine the desired type in the design.

Types of tourist village

Tourist villages are classified into different types according to the natural capabilities where they are located and built, as well as according to different maps and functions. Depending on the amount of features and attractions around, tourist villages may have accommodation or include only limited uses. After reviewing, we come to the conclusion that the village in question is of a design type.

Healthy towns and villages

Health tourism is a type of travel that, in addition to recreation, leisure and comfort, also brings health and medical care. Currently, medical tourism is one of the fastest growing sectors of the tourism industry in the world and is rapidly becoming a global industry.

All over the world, 7% of the total number of people who enter countries as tourists (international tourists) are those who have taken action to treat and use the natural gifts of that country. Areas of activity in the field of health tourism in the world include: Mineral spas (Health Spa), Weight management programs (Program Weight Management), Cosmetic and Plastic Surgery (Cosmetic and Plastic Surgery), Knee replacement surgery, ... (Knee replacement), Coronary By-Pass Surgery, Organ Transplant, Lasik Surgery, Dentistry and dental implantation, rehabilitation (Rehabilitation and Recuperation) and ....

Leading regions in the field of health tourism in the world (advanced in this field) can be Bahrain, Singapore, Cuba, Costa Rica, Hungary, Jordan, Lithuania, Malaysia, Thailand, Belgium, Poland, Turkey, Dubai, USA, South Africa and India. cited. (Zahedi, 17: 2006)

Coastal Village

Coastal villages are usually referred to as suburban areas that are formed and located in coastal areas that have the ability to attract tourists. Like the coastal village of Panormo in Greece, which was originally a small village but many factors have made it a tourist village. In coastal villages, the main factor in the formation is proximity or proximity to the beach (Shahraki, 2011: 29) that the village in design is close to this type of village in terms of some features.

Mountain village

Mountain villages that are formed in mountainous areas usually have the ability to create mountain resorts, and depending on the geographical and climatic conditions, activities such as skiing, snowboarding, mountaineering, rock climbing, hiking, snow climbing, snow climbing and hiking can be done. He planned the paraglider and so on. (Shahraki, 31: 2011)

Figure 16. Mountain village

Desert villages

Desert villages are formed in desert areas. These villages have residences for activities such as stargazing, sand safaris using special equipment, camel riding, ostrich riding, familiarity with herbal medicines and wildlife. The most important countries that have been active in locating and equipping desert areas to develop the tourism industry are the UAE and the deserts around Dubai. (Shahraki, 33: 2011) According to the mentioned cases, the type of village intended for design is almost of the type of coastal villages.

Tourism

The position and function of tourism in urban life

Probably the most places where tourism takes place are cities. Urbanization, which is the most complex human achievement today, has different physical, social, economic, cultural and anthropological dimensions. And the more a city is exposed to the use of tourists and tourists, the more it will be affected by the mentioned factors. Leisure, recreation and tourism are words that everyone has heard or used many times. For most of us, leisure is equivalent to leisure, but recreation refers to activities that are done to spend leisure time.

The word tourism often evokes in our minds travel, holidays and scenic sights. Travel to cities and use of public spaces for entertainment and leisure Tourists go to parks, cinemas, museums, shopping malls, gardens and various exhibitions, pilgrimages. Attracts places, buildings and monuments. There is a difference between spending leisure time for citizens and tourists and CIA tourists because spending leisure time for tourists requires travel, so the type of demand is also different. (Imran Zaveh, 2003: 90)

The structure of a city that focuses on tourism is different from an ordinary city. Establishment of accommodation and hotels, leisure centers, library facilities, catering and guidance in urban sightseeing centers and transportation should all be tailored to tourism relations. All these facilities are vessels for exchanging the culture of cities to tourists and from them to citizens. It is located in the area of the site for the design of the city hotel, which in terms of capacity is responsive to the clients of the complex and meets the needs of accommodation in this village. Also, in terms of the need for cultural spaces, despite the cinema with two halls and a city museum on the very close edge of this site is responsive, but it is suggested to use the axes that come from them to design the site to somehow this functional coordination in the site.

Ecotourism

As we know, man and nature are inseparable combinations. Man is born in nature, lives in nature and dies in nature. All natural features including natural forests, national wildlife parks, rivers and nature around them, especially waterfalls, mountains, natural springs, summers, hunting areas, fishing in the seas and lakes and their surroundings, habitats. Indigenous as well as natural caves determine the extent of nature tourism. The International Ecotourism Society defines ecotourism as follows:

It is a responsible trip to natural areas that, while protecting the environment, also brings health to local communities. Ecotourism is actually a journey to a natural area, a journey that is beneficial to local communities. A journey that has led man to a deep understanding of nature and the environment and to protect the protection of biodiversity. (Kiarostami, 2003: 33) Ecotourism is a combination of the words ecology and tourism, and in fact is the product of many challenges between extremists, fans of the use and unlimited use of nature and fans of rational and rational use of these resources, which in Persian means ecotourism and ecotourism. (Amirani, 2003: 4)

Ecotourism is a new phenomenon that represents only a part of the whole tourism industry and is based on purposeful travel to relatively natural areas for study, enjoyment, spiritual use of landscapes and any kind of contemporary or past cultural activity in these areas. In this dissertation, design according to the ecological approach is also in line with these criteria, so we will give a brief explanation in this regard.

The term became popular in 1965 and introduced four criteria for ecotourism.

1- Minimal negative impact on the environment.

2- Minimal negative impact on culture and maximum responsibility towards the culture of the host community.

3- Maximum economic benefit for the host community.

4- Maximum recreational satisfaction for the participation of tourists.

Today, on a global scale, it is accepted that ecotourism should be related to environmental protection and appropriate and long-term use of nature with minimal changes in the balance of the natural environment and harm (Mohammadi, 2003: 17)

Sustainable ecotourism

The basis of ecotourism is how tourists can visit natural and cultural resources to promote, raise awareness and spend their leisure time and enjoyment without having a negative impact on these resources. Conservation of recreational resources, along with productivity, is very sensitive. Sustainable management of recreational resources is synonymous with the term sustainable ecotourism.

Ecotourism is a type of tourist management in which the ecological system is protected. Protection and improvement of the natural environment and preparation of executive plans to promote natural beauty by developing green space and creating recreational areas in nature, such as picnic areas, camping, rest combined with multifunctional information systems to enhance the knowledge of users. Environmental protection will be one of the goals of the ecotourism development plan. The great recreational potential of forest and national parks, etc., and their special place in the tourism industry in the development of ecotourism are considered. (Shariatnejad, 1997: 14)

The main claim of ecotourism proponents is that ecotourism is based on stimulus strategies, practical strategies and preventive measures by activating and exploiting the sensitive natural areas themselves and promotes sustainable development. Because by doing so, it keeps such areas out of the scope of employment and threatening the aggressive development of the human environment (Behroozfar, 2003: 11).

One of the goals of this design is to protect the river, which is a sensitive natural area, from employment and the threat of aggressive human development operations. Therefore, according to the ecological approach to design, one of the proposed solutions is to design buildings with a delicate volume of mini-mall and low interference in the site. The concept of sustainable tourism and vakotourism is an acceptable and praiseworthy method that, while meeting the recreational needs of current tourists, must also take into account the fact that the affected areas also belong to future generations and must be used properly. He preserved them so that our children could also touch these precious gems.

Proper use of ecotourism or environmental tourism can have a positive effect on environmental protection in addition to socio-economic benefits. For example, the proper use of land causes the land to remain covered with its native plants, and in addition to numerous internal economic benefits, it attracts many tourists and leads to domestic and foreign investments in the country's infrastructure and ultimately sustainable development. (Mohammadi, 19: 2003)

Iran and tourist villages

In order to build tourist villages in environmental areas; Based on the agreement between the Cultural Heritage and Tourism Organization and the Environmental Protection Organization in 2001, a three-axis network of tourist villages has been identified. The base of this network was formed in three axes of Alborz, Zagros and Kavir with special features of these areas and tourists who want this climate. The Lut Desert passes through northwestern-southeastern Qatar, a route to Europe and Pakistan that is thought to have taken place. It is the best route for European tourists and has created special attractions for this group.

The Zagros axis is a route in the north-southern heights of Iran that passes through the provinces of Kurdistan, Kermanshah, Lorestan, Chaharmahal, Fars and Bushehr and includes the characteristics and charms of the life of indigenous tribes; A route that has special attractions even for domestic tourists. Alborz axis is a route in the east-west of Iran, which passes through the provinces of Azerbaijan, Gilan, Mazandaran, Golestan and Khorasan and currently has the largest share in domestic tourism in Iran and annually hosts more than 20 million people is a domestic tourist.

Figure 17. Iran and tourist villages

The desert axis includes the tourist villages of Qom Wandering Island, Taft Yazd and Mahan Kerman, and the Zagros axis includes the tourist villages of Shurab Lorestan, Koohrang Chaharmahal, Biston Kermanshah, Maharloo Fars and Shif Island of Bushehr.

Alborz axis has the tourist villages of Ashuradeh Island, Gaz Port and Turkmen Port in Golestan, Gisum Forest Park in Gilan, Islamic Island in East Azerbaijan and Neishabour in Khorasan to join the promise of tourist villages. (Rahshahr International Group, 2009: 75)

Presenting similar patterns and examples in the whole, component and micro level of the collection spaces

In the last step of the process of presenting the initial views of the design in relation to the subject of the design, it will be the turn to present similar patterns and examples of the total, part and micro-spaces predicted at the collection level. For this purpose, in this section, based on the implemented examples, an attempt is made to provide a preliminary perspective on the system of form and function of the components and wisdom of the envisaged spaces in the tourist village of Warsh.

Today, tourism has become one of the most important resources in the economy of countries, which has now grown more than other economic sectors. Which not only economically creates employment and income and ... It has become familiar with cultures, fighting unemployment, social corruption and increasing the welfare of society. So that it is so important in economic and social field. What economists call this industry ("invisible exports") (Saghaei, October 2009).

The importance of tourism in the present era is most dependent on its economic cycle, which has a high potential in the field of local and international economic dynamics, such as tourism consumption, public and private investment, as well as exports in the tourism industry in 2004 urban The equivalent of 5.9 percent was about $ 5.5 trillion. (Goli, 2008)

Tourism can be an important industry for countries that have not been successful in terms of production and exports worldwide. Although Iran has not had much success in the world tourism position, but it has rich resources in the field of tourism. According to the Cultural Heritage News Agency, Iran ranks 137th out of 183 countries in world tourism in 2009. This ranking confirms that Iran has not grown much in this field. Research has shown that countries that have reduced the government's dominance of tourism activities are in high demand in the industry without the participation of the private sector and government support, as well as other factors such as the security of life, the finances of foreign tourists, and infrastructure. And desirable welfare facilities in this field and regular and coordinated advertising. Tourism development is not possible. In Iran, the tourism industry needs macro and long-term planning, standardization and also the use of global trade in the localization of services and services of the tourism industry can be of great help to activists in this field. (Goli, 2008)

Iran should work in this field because it not only promotes economic development and income generation, but also creates a suitable infrastructure and enriches the culture of the society and preserves the environment due to the awakening of the people and their awareness of the value of their region's heritage and awareness. People are also increasing in this area. Among the countries that have had the most success in attracting tourists, Switzerland has been able to use its current and potential capacities and capabilities to have a superior position in this field. (Goli, 2008)

Therefore, the tourism industry is the third most important industry in the world after the oil and car industry, but it is still the most important industry. Because oil is the result of a natural process and is as perishable as the machine-building industry, and it is not possible to use them to paint a bright future, but the tourism industry is an industrial industry that is sustainable and whatever society develops, and brings machine life, this industry is important. Makes more visible. (Saghaei, October 2009)

During tourism, cities are considered by tourists that have several attractions or at least one of them, such as the existence of shrines, scientific, cultural and historical monuments, recreational and accommodation facilities, communication facilities and the existence of various markets and sales. The existence of urban tourism cannot be summarized only in the existence of attractions.

In a way that tourism as a product of the intertwining of various factors, each of which has a significant impact on the flow of tourism. In addition, the knowledge of tourism is also important in the field of urban tourism. Understanding tourism in relation to its nature and spatial pattern helps to adapt the local flow of tourism to an endogenous development trend, while only considering the profitability of tourism without knowledge can create many challenges for local residents. (Saghaei, October 2009).

Chapter III

General investigation

Tourism is the Persian equivalent of the word tourism in English, French and German, which is literally translated into Persian as tourism. It was then introduced to France and England, one of the meanings of which is to travel, which has become the source of tourism.

The term tourist, which has a common meaning in most living languages of the world, has been common since the nineteenth century. In the past, French aristocrats had to travel to complete their education and gain the necessary life experiences. At that time, they were called tourists. And later in France, the term was used for those who traveled to France for entertainment, leisure and travel.

Gradually, tourism entered other languages and the word tourist emerged. From that time on, tourism was referred to as some travel and travelers whose purpose was to rest and tourism and entertainment and acquaintance with the people and not to earn money and work. Tourism in each period was done with different goals that were appropriate to that period, for example, in some periods in order to get acquainted with the way of governing and life of the people. Tourism has been the same concept at different times but has had different names. For example: "Tourism is a set of phenomena and connections resulting from the interaction between capital tourists, host governments and communities, universities and non-governmental organizations, the three processes of attracting, transporting and receiving and controlling tourists and other visitors." (Papli Yazdi, Saghaei, 2006; p.12)

Of course, it must be said that tourism existed even in ancient times, but it was not used as a word tourism. As we know, in the past, few people enjoyed their leisure time, which was now religious, mostly in the form of pilgrims to religious places, and even in the time of kings, trips were made to acquaint the king with the different ways of governing and living. At different times of these travels, for example in the Middle Ages or BC or the Industrial Revolution until today, tourism has been done for various purposes, which is now referred to as tourism. (Kazemi, 2006, p. 7)

But the word tourism is defined by various individuals and organizations, including:

"Tourism is the activity of people who travel outside their usual place of residence for leisure, work and other reasons and stay there for a maximum of one consecutive year."

"Domestic (or indigenous) tourism": Persons residing in a country who travel for a maximum of twelve months to a place in their own country that is outside their normal living environment, and their main purpose of this trip is not to do something that ultimately received Wages to be visited from the country.

Tourism in general, due to its interdisciplinary nature, has the ability to have different attitudes, which has led to many definitions of it. In the initial definitions, more emphasis was placed on the distance, based on the distance they had from the place of residence. For example, the US national tourism commission had considered a distance of 50 miles, which includes trips other than business trips, because this definition was accepted as an economic and statistical quantity, but did not consider other aspects of tourism, such as supply. And he only paid attention to the demand side, and therefore tourism needs other definitions. Some of these different dimensions include geographical. The word "tourism" was first officially used by the United Nations in 1937, but the tourism industry is much older. It was used when a person traveled abroad for 24 hours. The word also included domestic travel and could even be extended to day trips.

King George III of Britain is known as the first tourist to spend his holidays on the beaches of Wimurtham whenever he fell ill. (tourist attractions website, 2006, support blog of Shivar Koohestan Tourism Institute) It was the oldest common form of tourism in Europe by the nobility of young princes to different places and to get acquainted with different ways of governing and living people. (Kazemi, 2006, p. 8)

BC when the Phoenicians traveled far and wide for trade by sea and land, or in the middle ages when most trips were made by merchants. But from the fourteenth to the seventeenth century they gained knowledge and experience, and since then universities such as oxford and Cambridge have paid for student travel grants. (Kazemi, 2006, p. 14). The first study of tourism in the United States was conducted by McMurray, Stephen Jones, and Robert Brown. They focused their studies and research on this issue, but we can also name people who were tourists and whose goal were exploration.

But in the Industrial Revolution, most of the trips were in groups, and most of the simple agricultural workers from rural areas went to cities to work in factories, where middle-income people still had little opportunity to travel. (Rezvani, 2002, p. 18)

Tourism in Iran, especially since the seventeenth century, the peak of the Safavid rule as an attractive country, attracted the attention of many European tourists. Which was the title of the beginning of the trip of foreigners to Iran, which is also mentioned in the pre-Islamic writings, which are before the trip of some Greeks and Romans to Iran. (Rezvani, 2002, p. 20).

During the reign of Reza Shah, by sending Iranian students to the West and expanding political and commercial relations, as well as the construction of European-style hotels in different parts of Iran, it was on the verge of becoming one of the tourist attractions and wrote travelogues about European travel to Iran. Such as the travelogue of Benjamin Todlai from Spain or tourists such as the travelogue of Peter Delavalle who visited Isfahan and the knight Sharon Chris who were the real missionaries of tourism in this land for world tourism can be seen. (Rezvani, 2002, pp. 189 and 190).

Today, tourism as an industry has positive effects that these effects include various economic, cultural, social, etc. dimensions that are economically significant to the extent that it covers the expenditure of about 7% of world capital and also causes Reduce delinquency and create employment and cross-cultural discovery. The tourism industry is important in two ways:

A) Provides people with familiarity with other cultures, races, ethnicities, lands, dialects, etc. Today, the economic aspect of this industry is more important. In 2008, according to estimates, the countries with the highest number of tourists in the world include France with 80 million foreign tourists, the most successful country in the top 10 countries in the world is France and the United States with 58 million Foreign tourists and Spain, China and Italy each ranked 57th, 53rd and 43 millionth, respectively, in third place in terms of tourism. In this ranking, the United Kingdom ranks sixth. For example, a quarter of French tourists travel to Paris to see the city. Or more Spanish tourists go to see its southern coast. Italy's income in 2008 was equal to

Iran's 2009 budget, and the Germans spent twice as much on tourism this year as Iran's 2009 budget.

One of the largest and most diverse sources in the world today is the tourism industry and in many countries is considered as the main source of national income. In this regard, our country, Iran, despite its natural and historical attractions, has been able to achieve success in this industry. Today, economists refer to it as an invisible export due to the increasing impact of the tourism industry on the economic and social development of countries. (Khairuddin, 2007, p. 8).

Figure 18. Importance of Tourism Infrastructure Development

According to forecasts, the revenue from the share of tourism in the world by 2014 will reach 736 billion per year. In addition to the financial consequences for the tourism industry, this industry in many places enriches the culture and awakens people to protect cultural heritage and increase information and social awareness. Among the various countries in the world that have made extensive efforts to develop the tourism industry and attract tourism, Switzerland is one of these countries that has achieved significant success in this field. According to the official statistics announced by the NAJA Immigration and Foreign Citizens Police, in 2009, 2272575 tourists entered the country through land, air and sea borders, of which 71% were foreign tourists by road, 27% by air and 0.7 tenths of a percent entered the country by sea and 0.1 percent by rail. Due to its valuable historical background, geographical location and other such factors, Iran has the ability to be located in a suitable position of tourist attractions in Asia and internationally. But with this description, despite having significant attractions, it does not have a good position in the tourism market. In the meantime, Iran has not had much growth and increase in relation to tourism since 2004, which in 2009 was ranked 137th among 183 countries.

 A comparative study of domestic and foreign tourists in the tourism industry of Iran shows the importance of this industry in the development of its domestic economy, so that according to statistics published by the tourism organization, the ratio of domestic to foreign tourists is 10 to 1. (Nobakht, Pirooz, 2008, p. 10).

In Iran, due to the existence of numerous historical monuments, there is a lot of potential for attracting foreign tourists, achieving this without creating the necessary infrastructure and providing the necessary facilities to attract tourists through the implementation of short-term, medium-term and long-term plans by tourism and tourism authorities will not be possible. (Goli, 2009)

The old context is the part of the city where several valuable historical-artistic or archeological works are located and represent an important and historical event or event, and its structure and configuration is such that the history of urban planning and architecture expresses it. The old texture of the city is a texture that has urban life and spirit and life flows in it. (Sobh Iran Independent Newspaper, 2007, Sanandaj Creation) Since its inception, human beings have made every effort to meet their needs and overcome obstacles in their path, and so far, they have succeeded in obtaining some gifts, the result of which has led to the emergence of culture and civilization. Old and historical contexts are able to depict this feature and show it in the form of buildings and components and elements of ancient and historical contexts. But these tissues have many problems, such as welfare-civil-social problems. However, some experts have offered a way to deal with these problems by changing and evolving in this context, but others are opposed to any change due to the cultural and historical importance of this sector. But regarding tourism in this part of the city, i.e. the old-historical context, it should be said that while accurate information about the existence of various historical attractions of our country at the international level can help attract tourists and generally develop the tourism industry. One of the most important provinces of the country, which can be considered as one of the important tourist spots with its many valuable historical buildings, is Semnan province, especially the city of Semnan. (Dehghannejad, 2010)

There are reservoirs, gates, castles, towers, mills, historical houses, etc. Most of the historical monuments and buildings of this city can be seen in the area of the old texture which is located in the center and south of this city. (Ahmadpanahi, 2009, Wikipedia site)

Figure 19. World tourism gradually recovers amid global uncertainty

The old texture of the city of Semnan, like the texture of the Islamic period, puts the three main elements of the mosque, the bazaar and the neighborhood together, and if today critics object to the texture and urban style of that time, it is because the design and texture of that time It was a time and place of that time. It seems that these elements can play an effective role in attracting tourists. But the question arises whether the necessary measures have been taken in this regard or not?

Or how can this city and its old context be turned into a tourist destination and be successful in the field of tourism? These are the questions that come to mind, but of course, if the infrastructure is planned and completed, and along with a look at tourism, improving the potential values of this context and its elements, this city can be considered as one of the attractive areas and a tourist destination. He said that there is a need for long-term and macro planning and the use of experience in this field can be of great help to activists in this field.

Not only has success been achieved in this field, but it has also reduced unemployment, created employment and awakened people and officials, and drawn their attention to this important issue and made their efforts to solve problems in this field as much as possible motivated. As we know, today the tourism industry is one of the most diverse and profitable industries among the industries in tourism, you can expect permanent profitable productivity at the usual cost, which is why this industry is considered a valuable asset.

According to figures provided by the World Tourism Organization, international tourism grew exponentially between 1950 and 2005, with economic growth rising from US $ 1.2 billion to more than US $ 622 billion, and tourism in 2007. It accounts for 3.10% of world GDP. In the same year, the job creation situation in this industry with about 234 million jobs was promoted to more than 8.2% of the total number of employed people in the world. The economic importance of the tourism and travel industry continued to increase with a large increase in the number of passengers from 25 million between 1950 and 783 million in 2006 with an average annual growth of 6.5 percent. (Goli, 2008)

In 2008, the countries with the highest number of tourists were France, the United States and Spain, each with 80, 58 and 57 million foreign tourists. For example, in 2008, Italy's income from tourism was exactly equal to Iran's 2009 budget, and the Germans spent twice as much on tourism this year as Iran's budget in 1988. (Honarparvar, 2010). According to WTO studies, the number of tourists is expected to reach one billion by 2010 and the tourism industry to reach 736 billion by 2014, while

Iran is said to be among the top 10 countries in terms of tourist attractions. The world is located.

In 2007, Iran was ranked 35th out of 178 countries and in 2009, it was ranked 137th out of 183 countries. Based on the comparison of statistics during the years 2007 to 2009 in Iran, it shows that the share of incoming tourists through land-based land compared to the total incoming tourists has increased from 58.6% in 2007 to 71.2% in 2009. This shows the upward trend of foreign tourists to the country through road borders, which indicates that the tourism industry in Iran needs a transformational view in the public and private system and creative to the tourism industry.

Due to having suitable tourism potentials and having a historical background, Iran can meet a wide range of needs and expectations of tourists, and a tourist with any purpose to travel to this country, can find the desired purpose and according to that the best way to get to know each land and its people is possible only through direct communication, and one of the characteristics of the Iranian people is hospitality and kindness, so with a little familiarity with all aspects of Iranian life, including thinking, He got acquainted with livelihood, customs, etc. easily. Tourist attractions in Iran include three parts: natural, cultural and religious. (Rezvani, 2002, pp. 216 and 217) that Iran is considered as one of the important poles of human civilization in the world due to its ancient civilization and being on the silk road and the connection between the East and the West. Iran has become involved, which has not only increased their importance but also increased the curiosity of tourists and researchers that the old cities of Iran with their traditional texture and historical monuments and ancient customs are among the historical and cultural attractions of this country. Recognizing, protecting and introducing them to the world will cause the prosperity of the tourism industry and the cultural promotion of Iranian society. (Rezvani, 2002, pp. 216 and 217)

Chapter IV

Theoretical foundations of research

Definition and concept of tourism

Tourism has been defined in different ways. The main and most scientific of which is as follows: In the technical definition by the World Tourism Organization (WTO), tourism is defined from different dimensions, based on the difference in approach to the place of visit, as follows:

Tourism is the activity of people who travel outside their usual place of residence for rest, work and other reasons, stay there for a maximum of one consecutive year. (Saghaei, Papli Yazdi, 2006, p. 12). But the definition proposed by Eric Cohen defines a tourist as: "A volunteer and temporary traveler who travels in the hope of enjoying a varied and fresh experience during a long and often unrepeatable journey." "(Armghan, 2007, pp. 24 and 25). According to swarboke; Tourism today is a tool for the development of developing countries, to the extent that Dukat dekudt refers to it as a passport for the development of countries. (Armghan, 2007, p. 25)

According to Casters; Tourism is a multidisciplinary science that will certainly not be perfected if it is developed without careful political analysis. (Armghan, 2007, p. 25)

In Persian dictionaries, tourism is defined as; Traveling and getting to know the world; Travel for fun and entertainment; And a journey in which the traveler goes to a destination and then returns to his place of residence. (Alwani, Deheshti, 1994, p. 18)

Tourists

A person who enters the country for legal reasons other than immigration and stays there for at least 24 hours and a maximum of one year. (Rezvani, 2002, pp. 17 and 18)

Literary definition of worn and old texture

The worn-out urban texture in English literature is equivalent to the term Inner city, which is largely the same as what we call the old texture of time. In the zoning model proposed by the Chicago School sociologists in 1920, Inner city is mainly defined by the following features: There he moves to the suburbs. 2) transformation zone; A gene interansition is a region whose inhabitants are either fleeting or members of the lowest social groups and stagnant and immobile economic groups (Azimi, 2009, p. 630).

From the point of view of the Cultural Heritage Organization, that part of the city that is about a hundred years old or more is called a historical context. (Tasian, 2004, p. 7) According to Dr. Mansouri, a member of the faculty of Tehran University, Baft has a unified set, of an integrated whole that should be recognized as a system. In a clearer definition, the context has a separate status and identity that has nothing to do with each historical monument, although historical monuments are the cells that make it up (Mansouri, 2004, p. 9).

The concept of old urban texture

The old texture is an interconnected set of urban components and elements, including residential units including dilapidated, restored and destroyed, valuable historical monuments, markets, facilities, grids, architectural form and special body that is the product of gradual and organic growth of the city. In historical periods, it is based on pre-industrial transportation technology and has a distinct construction in terms of function and appearance compared to new urban areas.

Therefore, it is not very compatible with the economic and social conditions of the present age, especially transportation technology, and horse access, narrow network of roads, lack of green space, training and burnout are among its major problems. In this part of the city, where the basic core of the city is formed, unique historical, cultural and artistic works called historical sources can be seen, which is an objective symbol of past civilization and a symbol of urban life administrative (Ministry of Housing and Urban Development) such sections (parts) inside the old texture have been named as the old texture, which is necessary to preserve and maintain it.

In this context, in addition to valuable cultural monuments and dilapidated buildings, there are also ruins, restorations and being destroyed, which in some cases have been turned into slums, sheds, etc., or have been left in a dilapidated and unusable condition. Therefore, they have no historical value, which is known as the old texture. Therefore, it is necessary to clean, renovate and improve such buildings. In addition to the above buildings, there are a number of new units with durable materials and suitable real estate value, the need to maintain them is not hidden from anyone.

Therefore, the old urban texture is a mosaic of different components and elements with various functions. The spatial order of these elements and functions creates a special structure for this part of the city that a one-sided approach makes it impossible to change the structure, so its organization requires a comprehensive approach to all its socio-economic, physical and physical dimensions. That is, a one-sided physical attitude, such as widening the passages, alone does not solve the problem of the old texture. The social and economic issues of the residents are also of special importance. That is, the context of tissue erosion should be considered in the social and economic conditions of its inhabitants, such as issues of ownership (inheritance, endowment), income, jobs, change of role and function, as well as the transfer of activities to new urban sectors such as markets, working capital and ... searched. (Rahnamaei, 1995, p. 53)

Definition of tissue types

A) Historical context; Historical texture is the areas located in the old parts of cities that before the beginning of the present century, i.e. the beginning of new urbanization in Iran, formed the surface of the city. Such textures are now located in the center of cities. They have a special position in the city and a relatively large level and strong performance (metropolitan and regional and national scales) has added to its importance. Traditional urban markets as trade centers and other important buildings such as religious centers are located in these contexts. (Romina Thesis, Supervision by Yousef Ali Ziari, 2005, p. 30)

B) old texture; The old texture also began to form around the original core of cities (i.e. historical and continuous texture). This part of the city, which was formed in the interval between the transition from quiet urbanization to rapid urbanization, is neither historical nor very new. Even its spatial organization is something between a historical and a new context. It can be imagined that the body of Iranian cities in the first three decades of the fourteenth century (AD) was composed of historical and ancient context. (Rumina Thesis, 2005, p. 30)

C) New texture; In the early 1940s, the growth of Iranian cities began faster than in previous decades and around the old texture. There was historical and religious, the city was expanding towards them as well. (Romina Thesis, 2005, p. 30)

Theories of tourism (in the world and in Iran)

Numerous experts on tourism have proposed theories, some of which are mentioned here, some of which we will discuss in this section. People like Bradon, Ekranes, Getz believe that tourism is out of the one-dimensional and pure state and has put new perspectives on social and natural issues. In fact, if tourism does not align with other social issues and ecological regulations, the program an adjustment will eventually lead to failure. (Armaghan, 2007, pp. 206, 207 and 208)

Getz also considers tourism and planning in it as a process based on research and evaluation and based on optimal conditions in creating a relationship between tourism, welfare and environmental protection. But from an idealistic point of view, planning for tourism is required to be consistent with economic and social interests at all levels, and tourism planning is not merely a formulation of plans and plans for the future. But how to do it without damage. And tourism works in the direction of an open economy. In the framework of which, the goals should be clear, but according to Ekranz, three elements from different angles of tourism, which are:

1- Satisfaction of tourism.

2- Environmental protection;

3- Encouraging tourism operators and innovators. (Armghan, 2007, p. 25)

Eric Cohen in his study of tourism on social life points to four types:

1- Public tourism; They travel to relieve fatigue, which has little effect in the region.

2- Individual tourism; That the tourist has more freedom of action and because he travels alone.

3- Nojo Tourism; The rejection of those tourists travels alone and tries to find a new way to connect with the host community, which is very effective.

4- Easy tourism; Leave your home and start a new life in the host community, looking for new issues and having the greatest impact on the host community. (Armghan, 2007, p. 26)

But Hamid Zargham says; Tourism in the 21st century in the context of technological innovations and new managerial functions on the one hand and the total domination of capitalism along with the formation of the world economy and the weakening of political borders cause many changes in geographical spaces. Tourism in the deconstruction of postmodernism It is an important social reality that is viewed mainly from the perspective of cultural balance and developmental balances (social justice). (Zargham, 1997, p. 390).

And Mohammad Reza Shafaghi also says; The performance and impact of tourism in geographical spaces is beyond the two-dimensional concept in the field of culture and economy and forms many perspectives. The study of the effects of tourism from different perspectives has crystallized in a comprehensive whole in economic, social, cultural and environmental dimensions and has formed a multidimensional approach. Geographical spaces such as tourism destinations and goals are constantly changing (Shafaqi, 1997, p. 227).

Attitudes and views about the old texture

Museum Attitude: This attitude is based on the protection of cultural heritage and does not accept intervention in historical or ancient contexts except in order to protect them. In this attitude, preserving the identity and heritage of ancestors is preferred to the necessities of contemporary life. As a result, the inhabitants of such contexts are forced to leave the context due to the lack of appropriate changes in contemporary daily life, as well as the shortcomings and inadequacies of urban infrastructure rooted in such a vision. Following the disintegration of the social fabric, the physical fabric is also subject to destruction.

The attitude of a museum in a logical sequence threatens the opportunities of life, and in some cases the next generations are condemned to live with the values of others, thus living in spaces that have not arisen from the necessities of their lives. In this view, the effort is focused on preserving the physical-spatial aspects of the old tissue and neglecting the other vital aspects of the tissue.

Cellular approach: The indisputable principle of the museum approach is that an entity, no matter how complex, can be divided into components and by studying the existence and behavior of its components, the original existence can be identified. The cellular approach in the Renaissance began with concepts such as reductionism and determinism and culminated in mechanical ideas and the industrial revolution. In general, in this view, especially in the period of the industrial revolution, the existing historical monuments and the body of valuable old buildings of the city are considered as a tool for economic empowerment of the city. The main motivation for their protection and restoration is the modernization of the economic structure of the city.

In such measures, urban restoration is carried out on the condition of guaranteeing economic benefits and achieving the improvement of some sections in the social conditions of the citizens, and any possibility that seems useful is used. The field of planning for urban restoration is more about renovation and renovation - improvement of the old built space of the city and less about improving and reviving it, with a view to the environment and human environment. The starting point or starting point of these measures is the revitalization and improvement of parts of the city that have lost their efficiency due to physical exhaustion and lack of access to modern technical-welfare service networks, and for this reason, residents, employees and special clients are interested in them. They do not live in such urban structures.

Restoration works pay maximum attention to the quality of construction works and consider the renovated or improved appearance of buildings and textures as a value that can be used by everyone. In this way, the idea is not bound to the continuity of cultural-emotional-social values, residents and employees (and not owners) are often not left out and interfered in the research, studies and decisions that design urban remodeling. (Consulting Engineers of Ivan Naghsh Jahan, 2008, Volume 2, Page 8)

The basic characteristics of "urban restoration with the aim of urban empowerment" or cellular method are:

- ❖ Relative maintenance of the physical body of the building.
- ❖ Strengthening restoration positions or places.
- ❖ Integration of restored texture with adjacent spaces.
- ❖ Independence of urban planning actions from citizens and the social fabric of the city.
- ❖ Limitations in the scientific and epistemological space of the project.

Due to their local nature, cellular encounters with the ancient-historical context may disrupt the functional order of the city and intensify the disturbances in the spatial organization of the city due to the negligence of the studies and the major economic aspect of the project. (Avan Naghsh Jahan consulting engineers, 2008, 2, 9)

Organic attitude: Organic thinking style emphasizes new concepts such as generality, system, hierarchy and dynamics of biological phenomena. In the organic attitude, the old texture and its elements are examined systematically in relation to each other and in relation to external spaces. The state of the old-historical closed system is removed and through the correct explanation and organization of their relationship with the new parts of the city, it achieves its high structural position in the city. The attitude of improvement, renovation, revitalization and psychology of the historical context should ensure the connection between the past and present life of the city and determine the direction of its future development. In this view, urban restoration with the aim of sustainable development is very similar to the restoration of historical textures in an organic way. The main feature of urban restoration in this method is:

- ❖ Maintenance of physical corpses.
- ❖ Unification of city living conditions.
- ❖ Reintegration of old urban centers and structures in the environment.
- ❖ Involvement of citizens in urban restoration.
- ❖ The topic of urban improvement in multidisciplinary circles.

An attempt is made to direct all the developmental processes or processes of the urban researcher in the imagined direction (resulting from the goals of the program-physical). From small to large, such a movement (creation of the whole) must be considered. In the organic approach, historical and architectural values should be considered as part of the general values of urban life in live interaction with the city and citizens, and physical protection should be combined with social protection and public interest. In this approach, in regeneration, apart from the appearance aspects, it also focuses on functional features and elements and their role in the new urban structure is a city. (Ivan Naghsh Jahan Consulting Engineers, 2008, Volume 2, Page 11)

The process of intervention in urban tissues as an intervention in the field of worn, ruined and disordered urban tissue began with the beginning of modernization in the spatial-physical organization of the old tissue after 1925. During this period, the government of the time, by making changes in the production organization, provides the ground for the physical changes of the city. The approval of the law to widen the development of roads and streets in 1933 and the amendment of the law to widen and develop roads in 1941, the construction of streets and squares with a different structure from the structure of the old city, the destruction of neighborhoods, etc. were among the measures of this period. (Ivan Naghsh Jahan Consulting Engineers, 2008, 2, 11)

Schools and theories about ancient texture

Interference in the old context requires its objective knowledge of the geographical space of the city and its relative distinction from the middle and new urban contexts. For this reason, understanding this context is rooted in theories of urban construction. As a result, determining the scope of the old texture, both physically and physically, and its functional role, requires a general study of theories of city construction. (Shokouei, 1998, p. 151)

And theories about the old texture:

The theory of Eugene Violet Sudo (1814-1879) Engenevidlet le-duc: Ludo's theory on the revitalization of historical or ancient monuments in the context is summarized in the following points:

A) All parts or elements that were added to the building at times after the original date should be removed from the body of the building and by doing the necessary work, it should be reflected in its original form with respect to the unity and purity of style.

B) If the building loses part of its body due to destruction caused by human factors or desires, and even when the building is not completed in its physical unity and is essentially unfinished, the restorer must complete the building and Make it the original body it should be.

C) In performing any of these interventions on the existence of the body of the building, the restorer must assume himself as the place of the main builder and according to the knowledge and understanding of what he could have done at that time "logical model" representing the building in the figure the original should arrange and compile it and use it in the moment of intervention in the construction of the building and give the building the same shape and body that the original builder, if it still existed, would create. (Romina, 2005, p. 38)

Theory of John Ruskin (1900-1719) John Ruskin: Great English writer and social and artistic believer, Ruskin directly on restoration, architecture theories or guidelines that can be included in the restoration of architecture. He has not left, but his personality and cultural position are famous in England and in Europe. He believes that recognizing an ancient monument is a historical document and should be avoided as much as possible in order to eliminate uncoordinated parts or add new parts and strengthen the unstable parts. And dangerous action should not be taken. (Flamaki, 2001, p. 15)

But there are three types of attitudes that are more related to the old context of cities between planned officials, urban planners and architects, which are:

Conservative Theory; In this view of phenomena such as buildings, ancient and historical textures, there is a kind of ritual thinking. Followers of this view believe that any intervention in the status quo should be avoided as much as possible. John Ruskin says that recognizing an ancient monument is a historical document, and that no action should be taken to remove uncoordinated sections or to add historical sections to strengthen unstable and dangerous sections. (Talesh Consulting Engineers, 1994, 20)

Radical theory: In this theory, to intervene in the old textures of cities, they prescribe the transformation or complete renewal of the old textures while preserving valuable cultural artifacts, and demolition and renovation are proposed as the only suggested way. (Talesh Consulting Engineers, 1994, p. 20)

Rational Theory: In this group, restoration is conditional on revitalization and renewal in old tissues. Based on which different stages are suggested for intervention in historical centers.

1- Buildings that must be protected.
2- Buildings that can actualize life and survival by changing their body and function.
3- Buildings that need to be repaired.
4- Buildings that need to be demolished and renovated. (Talesh Consulting Engineers, 1994, p. 20)

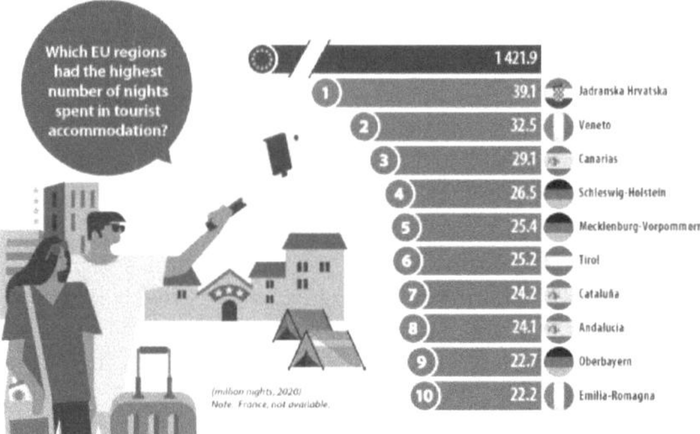

Figure 20. Tourism, in a statistical context, refers to the activity of visitors taking a trip to a destination outside their usual environment, for less than a year. It is important to note that this definition is wider than the common everyday definition, insofar as it encompasses not only private leisure trips but also visits to family and friends, as well as business trips.

World Tourism Industry

According to the World Tourism Organization, tourism is one of the most dynamic economic sectors in the world, which in the next few years will be at the top of the world's existing industries in terms of revenue and will account for a large part of international trade. The tourism boom improves the economic indicators such as production and employment in the agricultural and industrial sectors by increasing the production of food and industrial products needed by tourists and the service sector through the boom in the hotel, restaurant, transport and trade sectors. And for this reason, it acts as a development sector in the country's economy. (Damghanian, 2007, p. 25)

According to available statistics, the total expenditures made by tourists in destination countries (tourist income for destination countries) is about 1.2 times the total annual income of OPEC member countries from oil sales, which shows the importance of the tourism industry more than oil in generating income; Given the importance of the tourism industry in economic prosperity, there is global competition between different countries to attract tourists.

According to statistics published by the World Tourism Organization in 2002 (the number of tourists in the world was 675.1 million, of which the United States with 114.97, Europe with 399.8, Asia with 131.3, and Africa with 1 (29 million tourists), but in 2004, 777.5 million tourists visited the world, generating $ 743 billion in revenue for destination countries. France, Spain, the United States and China are among the most attractive countries for tourists, so that 27.8% of all tourists in the world have visited these four countries. Also, tourism revenue for the three countries of Spain, the United States and China in 2004 was about 190 billion dollars. (Damghanian, 2007, p. 26). According to statistics in 2008, the countries with the highest number of tourists in the world, including France with 80 million, the United States with 58 million and Spain with 57 million tourists, which ranked first to third, and Mexico with 22 million foreign tourists.

According to statistics published by the World Tourism Organization, by 2020 the number of tourists in the United States is 282.3, Europe 717, East Asia 397.2, Oceania 68.5, Middle East 18.8, South Asia and Africa 77.3 million. So that Europe with 7.7 and South Asia with 18.8 million people had the highest and lowest number of tourists, respectively. (Armghan, 2007, p. 178)

As for the list of major tourist destinations in the world, which was registered by the World Heritage Convention in 1987:

Table 1. List of major global tourist centers and hubs

Relevant area names and descriptions	country name	Row
The historic center of Rome, the paintings on the stone and the Alcomonica, the church and the monastery of Santa del Grazi with the Last Supper by Leonardo da Vinci, the historic center of Florence, Venice and its lake.	Italy	1
Mosque, Alhambra and Jenner Alif, Granada, Cathedral, Burgos, Monastery and Escorial Conservation, Madrid, Altamira Cave,	Spain	2

Segovia's Old Town and Canal, Santiago de Compostela's Old Town, Avila's Old Town with churches outside the Fence, and its fences.		
Mont Saint Michel and its bay, Palace and Park of Versailles, decorated caves of the Valley of the Wazir, Roman monuments, Church of the Seine, Victory Room	France	3
British Isles, North White Hills, Downs & Yorkshire, Wales	England	4
The Three Pyramids, Memphis and its tombs, the pyramids of Giza, new monuments from Abu Simbel to Fila, Islamic Cairo, Abu Manna.	Egypt	5
Sagarmatha National Park, Kathmandu Valley and Royal Chitwan National Park.	Nepal	6
St. Galle Monastery, St. John the Baptist Church, Mostier, Bern Old Town.	Swiss	7
Historic areas of Istanbul, Cappadocia Protected Stone National and Local Park, Grand Mosque and Diorighi Hospital, Hatosha, Nimrud Dagh.	Turkey	8

Source: Rezvani, 2002, p. 174

Tourism and Tourist Priorities in Iran: Special tourism facilities in Iran as well as the values governing the Iranian society cause our country to naturally attract a special type of tourists that can be considered according to the priorities Classified tourism in Iran. Of course, this classification should be adjusted and modified with the developments in the international and Iranian community. (Damghanian, 2007, p. 25)

Table 2. Tourism Priorities in Iran

Pilgrimages to holy shrines, mosques, temples and tombs of the great	Mystical pilgrimage	1
Visiting antiquities and visiting human gatherings, especially Iranian natives	Social history	2
Familiarity with literary manifestations and scientific standards of Iran and holding world conferences	Literary	3
Spending leisure time in the nature of the north, south, east, west and center of Iran	Natural recreation	4

Source: (Damghanian, 2007, p. 25)

Table 3. Priority of tourists in Iran

All Iranians who want to maintain their previous relations with their homeland	Iranians living abroad	1
Sheikhdoms of the Persian Gulf and all Muslims in Turkey, Iraq and Pakistan	Muslim neighbors	2
Iran's northern neighbors and nearby countries such as India and Saudi Arabia	Other neighbors	3
All those who are interested in Iranian attractions that have tourist intentions	Other intellectuals	4

Source: (Damghanian, 2007, p. 25)

According to various natural, historical, social and mystical attractions that can create favorable conditions for tourism, tourism can also be categorized as follows:

Table 4. Types of tourist places

Relative restrictions	Relative facilities	Prominent tourism hubs	Type of tourism
Procurement	Bus, hospitable	Hot water of neighborhoods, Jamkaran, Sarein Ardabil, Masouleh	Local
facilities	National Airport, 3-star hotel	Mashhad, Kerman, Gilan, Mazandaran, Kermanshah	National
advertisements	Regional airport, 4-star hotel	Kish, Qom, Golestan, Urmia	regional
Security	International Airport, 5-star hotel	Shiraz, Isfahan, Tehran, Hamedan	international

Source: (Damghanian, 2007, p. 25)

Global experiences with ancient textures

Most of the experiences in organizing the old texture of cities go back to England and Italy. The customs and habits of the British are different from other European countries such as France, which has a conservative spirit and preservation of national traditions is one of the prominent features of English culture. Using the laws, it has interfered in the old texture. For example, the city of Bath in England is one of the old cities whose collection of historical monuments has a great heritage value. In general, the set of actions that take place in the old texture of this city. Taken as follows: (Rasooli, 1997, p. 316).

    i. Reconstruction and renovation based on commercial and economic activities for greater profitability by attracting tourists.

    ii. Examining the old texture on two levels, both at the level of the building and at the level of the complex, which can be effective in the continuity of urban space, densities, access and continuity of the old texture.

   iii. People's participation in the reconstruction and transformation of the building and the performance or intervention of the property owner.

   iv. Tax exemptions, bank loans and facilities.

    v. Prevent land brokerage and speculation in the old context.

   2. Valuation of old buildings and textures based on criteria such as age, style, materials and performance. (Rasooli, 1997, p. 316)

Italy: The port city of Ravana is located in northeastern Italy and on the beautiful shores of the Adriatic Sea, which is not only a commercial port but also a historic city dating back to the fifth century AD and the era of the Roman Empire. Urban life in the historical context of Ravana is still going on and from aristocratic areas and tips on the protection and restoration of the historical context done include: 1) Implementation of urban unit management. 2) Emphasis on maintaining the flow of civic life in the historical context. 3) Prevent horizontal construction in cities. 4) The interaction of the municipality and the people living in the old context. 5) Raising the level of awareness and knowledge of the people regarding the revival of the historical context of the city, which in this city, by carefully studying the city's space organization, has determined the main thoroughfares and concentrated the most important economic and cultural activities there. Change its structure and passages. And of course, this organization is not without study and planning, and the perfect balance is noticeable. (Romina, 2005, p. 48)

Iranian experiences

The beginning of Qajar rule in Iran coincided with the Industrial Revolution and the arrival of foreign capitalists, which caused changes and transformations in the old textures that were completed during the Safavid period, and due to its long history.

Iran's urbanization has old textures in the heart of urban centers, which has many problems that have made urban planners think to pay more attention to it. So that this attention has caused changes in the old texture. (Erumina, 2005, p. 57)

Yazd texture: that considering the old structure of Yazd city which is the economic and political backbone of the city and the creation of new streets and new structure in this city that had problems 1) that never old and new texture with They were not compatible with each other.2) The boom of the old texture economy. 3) Loss of market power in front of the street. 4) The difference between the price of land and its goodwill in the old texture compared to the new texture. 5) Approximate destruction and erosion of old tissue. (Erumina, 2005, p. 57)

Experience of District 12 of Tehran: In this area, the old neighborhoods of Tehran have been improved and renovated in order to improve the quality of life of the people and its deterioration, and also the residents and private sectors have been encouraged to improve and renovate the old texture. (Romina, 2005, p. 57)

If we take an overview and want to compare the trade of Iran with European countries, we will find that Iran has failed in this regard. Because the experiences of Iranian cities show the deterioration of the old texture and the effect of external factors in changing this type of texture, and it can be said that in Iran, the old textures have been attacked by Western countries and intensify the gradual decline and wear and tear. Increasingly, they have become the core of their cities, but European countries, in addition to preserving the old-historical context by preserving valuable cultural elements as well as preserving the past identity of their historical context, have continued this part of the city and even their old texture is home to affluent people. It is high and in general it is a neighborhood of aristocrats and even the natives do not want to leave this texture, while in Iran due to the problems not only the natives do not want to live in this texture but also the neighborhood of low-income people and causes it to change (Farzam, 1997, p. 64)

Iran's position in world tourism

The Islamic Republic of Iran is one of the top 9 countries in the world in terms of buildings and historical monuments and is among the top 10 countries in the world in terms of ecotourism attractions. (Armaghan, 2007, p. 236)

However, in proportion to these capabilities, it should benefit from the $ 400 billion tourism revenue and its share should be at least 5% of this amount of revenue. But not only is Iran's income not this much, but Iran's annual income is much lower and about 5 percent of the share of tourism and is ranked 92nd in the world, which is unfortunate to mention and the need to study and solve problems in this area. We need careful and coherent planning to solve these problems. This failure is due to the implementation of protectionist policies and the adoption of laws in line with the establishment of a sustainable system of tourists and tourism. According to the 20-year vision, the government should attract 20 million tourists by 2020. But considering these problems and obstacles, achieving it is a bit ambitious. (Armaghan, 2007, p. 267)

But according to Iran's potentials, with sufficient knowledge and awareness of the conditions and a proper planning, we can succeed. At the same time, we should not forget that tourism has become the third most lucrative industry in the world today, after the oil and car industry, so we must pay attention to the tourism industry.

Because it does not need raw materials, it does not want heavy capital, and by creating investment and infrastructure, we will not only have an eternal and lifelong income for the country, but it will also lead to a fair distribution of income, reduce unemployment, and social corruption. In many countries of the world, tourism is the main source of income for governments. Countries that have oil, the oil industry accounts for a small percentage of their income and their main benefits are through foreign exchange, which enters their country through tourism. (Rezvani, 2002, pp. 17 and 18)

Old texture of Iranian cities

Most of the important and large cities of Iran are traditional cities. One of their important and open features is the formation of the city from two parts, old and new. In general, that part of the cities of Iran that until the early 20th century. They can be called the old texture and refers to the time when the impact of industrialization or urban change in the Pahlavi era (1924-1925) did not begin. The city of Iran is divided into three parts, inner and middle, which is called the old texture, which has slowly grown and formed over hundreds of years, and the outer belt of the city, which is the new urban texture, has been growing rapidly. Basically, the feature that the city is old and new and separable from each other is one of the characteristics of Islamic cities. The new urban part of Iran is also known as Doran Build and Sell. The size of this new part is usually several times the size of the old part. (Wall of Eivazi, 2004, p. 59)

Urban Development

In developed countries; Urban development and revival and attention to historical contexts have a very long history. In this way, logical plans and opinions have always existed and various discussions and theories have been constantly presented, which time is the main factor in the implementation of such measures in connection with the plan and plays a fundamental and important role. The timing of such projects in the old context should not be long, as stopping work in such environments will lead to the frustration of the people and private sector investors in it.

The faster and more fruitful the work, the more the beginning of the next stages will be welcomed and the people will be encouraged to participate and modernize. Functional diversification and reorganization in urban life historical textures should become an attractive place in terms of facilities and landscape diversity, in the process of which the return on investment will definitely be provided and the way to plan and continue the revitalization of textures. Smooth your history. Efforts to revitalize the historic neighborhoods of the city should be done within the framework of a sensitive context and environment. In this case, the processes of protection, preservation, and improvement of historic neighborhoods in the process of revitalization become a vibrant, active and dynamic environment. (Erumina, 2005, p. 36)

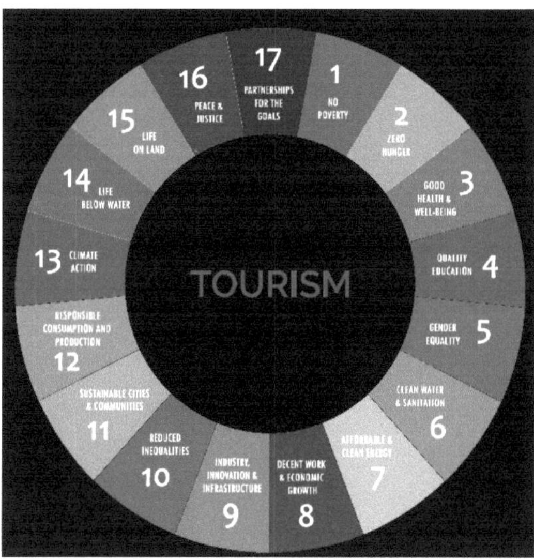

Figure 21. Visualizing the Countries Most Reliant on Tourism

Types of tourism

Tourism has been classified in different ways, the most important of which are as follows:

A) focused and extended; Centralized tourism usually requires special facilities such as swimming pool, ski slope, sightseeing tour, camping, cycling, visiting cultural monuments, etc., but extensive tourism does not require facilities and equipment. Such as mountaineering, hunting, or little need for facilities. Such as desert tourism, fishing, horseback riding, watching animals in nature, etc. (Rezvani, 2002, p. 18)

B) Inland and overseas; Inland tourism: It is tourism that is a recreational-tourist trip within the borders of a country and tourism in other countries is called overseas tourism. It should be noted that in today's world, the overseas tourism industry is the second most lucrative industry in the world. (Rezvani, 2002, p. 18)

C) Tourism based on the purpose of travel: Vance Smith divides tourism into 6 types; Ethnic, historical, recreational, artistic, naturalistic and occupational tourism. Of course, this classification is not immutable, because in most cases, tourists have a combination of these goals in mind and travel with these multiple goals. (Alvani, Deheshti, 1994, pp. 108 and 110)

D) Summer and winter tourism: Favorable and suitable natural conditions in Kana other tourist attractions cause the prosperity of summer tourism, which includes sunny days, suitable temperature period, sea water temperature, roughness and water storage, relative humidity and Vegetation. Here the duration of the bright sun is determined in a certain period. The average daily temperature is from 15 to 20 degrees Celsius in summer. (Shokouei, 1975, 28)
But factors that boost winter tourism include good winter temperatures, hours of sunshine, wind speed, snow cover, bumps, vegetation, snow height, and how long the snow stays on the ground, among others. There are factors that affect winter tourism. (Monshizadeh, 1997, 85)

E) Special and general tourism; Special tourism refers to a specific case of activities such as mountaineering, cultural heritage visits, fishing, etc., and mainly have a more scientific view of the environment, and tourists with the least facilities and equipment seek Is your interests. In public tourism, the tourist leaves home with the intention of spending his leisure time, and his main goal is to travel and have fun. (Rezvani, 2002, 18)

17-2 Tourist attractions

In general, tourist attractions can be divided into three groups: natural, human and cultural:

A) Natural attractions: Numerous mountains as natural attractions that are the best place for mountaineering and skiing enthusiasts and diverse plains, forests, and amazing caves are the most beautiful phenomena in the world that the view inside the cave is very original. And every year it attracts a large number of people that some caves, even with different forms of sediments, have created a special beauty inside. Seasonal deserts and lakes as well as hot springs, beautiful waterfalls each have their own charm. Which are visited by thousands of tourists every year. Iran is not without these attractions and can receive many tourists every year. (Armaghan, 2007, p. 223)

B) human and cultural attractions; Iran because of its rich civilization and culture and being on the Silk Road that connected the East and the West of the world in the past, and because throughout its ancient history, world governments and just empires have flourished. Seen for itself, it is full of historical monuments and relics, customs, works of art and colorful handicrafts. Among these attractions, we can mention historical attractions, treasures, handicrafts, customs and traditions, religious attractions. Cultural tourism and ecotourism programs are a priority for the cultural heritage and tourism organization, because of the existence of neighboring countries and the existence of many holy places in the country, this tourism must be managed. Forecast (WTO) says even the lowest tourist country in other words, Iran's real share of international tourism revenue in 2020 should be at least $ 50 billion, based on its natural geography and historical and cultural location. (Armaghan, 2007, 224 and 225)

18-2 Tourism and Sustainable Development: Sustainable tourism cannot be separated from the concept of sustainable development. In this regard, the dimensions that we can examine for sustainable development include:

1- Ecological dimension; Protecting the natural environment of tourist areas is a prerequisite for sustainable tourism. Therefore, in this regard, we must consider strategies related to the protection of environmental resources and reduce the spread of pollution in various fields and the preservation of the site.

2- Economic dimension; Activities in the field of tourism should maximize the added value of the area.

3- Socio-cultural dimension: All tourism planning processes (such as reviewing elements, analysis and formulation of policies, etc.) should be in the direction of cultural exchanges and community satisfaction. In the first place, tourist areas are not to provide services to tourists, but the neighborhood is inhabited by locals.

Then, first of all, it should improve the employment and living standards of the local people, and at the same time, their ideological and cultural moral principles should not be damaged. 4) organizational dimension; In planning for tourism, all organizations that deal with tourism in some way should be considered and our approach in this regard should be considered from the bottom up. 5) individual dimension; Meeting the demand and consumption habits of tourists has a great impact on activities in this field, and then it is important to pay attention to travel culture and tourist demand. (Armaghan, 2007, pp. 94 and 95)

But the characteristics that a sustainable tourism can have: 1) Tourism must be planned in a sustainable and stable manner so that all economic systems, environmental and natural resources of the desired location are supported. 2) In the implementation of tourism programs, equality and justice in the distribution of income between the host community and those involved in this field. 3) Research on the nature of tourism should be made available to all members of the community. 4) In all stages of tourism development, there should always be controls in all matters so that officials are confident that local people enjoy it fairly. 5) And in the development of tourism, there must be a reasonable relationship between different economic, social and natural conditions. 6) Different organizations should adhere to ethical principles and value the environment and the host community for the leadership practices of the community, but the goals that exist in this area and include: 1) Improving the quality of life of the host community. 2) Observance of equality between two generations within a generation. 3) Maintaining the quality of the environment by maintaining the environmental system. 4) Maintaining cultural integrity and cohesion and solidarity between communities. 5) Creating facilities and facilities so that visitors can gain experience. (Armaghan, 2007, pp. 97 and 98)

Social and service barriers, implemented in the tourism sector

There are obstacles in the development of tourism in the country, some of which are mentioned.

1) One of these factors is the low level of public awareness about tourism and tourist attractions in the country and not taking the resulting income seriously.

2) Increasing the cost of living causes families to sometimes think about eliminating travel expenses, as well as the lack of welfare and medical facilities and the lack of proper planning with them.

3) Another is that we are not as familiar with the principle of world trade as we should be, as well as legal and administrative barriers such as cumbersome customs regulations and visa issuance.

4) Lack of suitable facilities for hospitality and the need for expertise of travel agency managers as one of the indicators that the weakness of such services and neglect of tourism insurance in this area.

5) bottlenecks in the intra-city transportation system.

6) Weakness of facilities and facilities for entry and exit of foreign nationals, etc. (Damghanian, 2007, p. 25)

Many historical places have not been able to attract tourists for various reasons. These places have not been advertised as they should be. And in terms of care and maintenance, not enough activities have been done. Some have also been destroyed. There are not even enough information brochures about them (Damghanian, capabilities and tourism development strategies, pp. 25, 2007).

Socio-economic effects of tourism (tourism and job creation): If tourism is appropriate for proper planning and active management, as well as the protection of national, local, regional capabilities, can have positive effects. 2007, pp. 158 and 159) 1) Increasing the country's income due to the profitability of tourism. 2) Expansion of employment in different sectors and economic development of the country 3) Utilization of local resources and development of these resources 4) Development of the region due to investment in the potentials and capabilities of the region such as natural, cultural, social, historical and Etc. and create balance in the region. 5) Integration within the boundaries of the study area and knowledge and awareness of ethnicities and nationalities within the study area. (Which currently employs more than 200 million people worldwide in tourism) accounts for about 11% of total global employment, which would increase if tourism develops.

Tourism is now a lucrative industry, boosting the economy of the society and playing an important role in the relationship between nations and tribes in the world, which is an opportunity for tourism. It is an industrial tourism that creates employment opportunities for various jobs. For this reason, it is necessary to promote tourist attractions outside the borders in a favorable way in order to provide the ground for foreign tourists to travel and create healthy and fast employment in various fields such as transportation, food, and hotel management. Includes banking, insurance, etc. (Armghan, 2007, pp. 158 and 159)

Table 5. Positive and negative cultural and social effects of tourism

Negative effects	Positive Effects	Factor	Row
Changes in traditional-artistic activities, hustle and bustle in traditional logic	Increase support for traditional culture and display ethnic identity	Culture in attracting tourists	1
Prevalence of the disease and increase of negative manifestations	Eliminate negative manifestations and create social opportunities	Direct contact of tourists and locals	2

Increasing social inequality and conflict	Reducing economic and social inequality	Changes in the economic structure	3
Excessive pressure on recreational areas	Development of tourism spaces	Development of tourism organization	4
Population growth and crime	Increase health and service facilities	Population increase	5

Source: Armaghan, 2007, pp. 154.

Dry cities of Iran and its tourist capabilities

Due to the acute climatic conditions in these cities and the role of humans in shaping the passages, houses, open spaces and generally the old part of cities. In addition to these factors, the culture of the society and the historical-social aspects have caused the city to be compacted and the passages and houses to be safe. In these cities, the round wall of the tower and fortifications encloses the complex and the element of the citadel of the city has blocked the government center and the city court. The main elements of the cities are mosque, school, caravanserai, and reliance. The conditions of the arid regions of Iran are very important and attract countless tourists. (Armaghan, 2007, 125) As for cultural and historical-memorial tourism in the field of these contexts, this section includes tourists who are mainly educated and travel there to get acquainted with the history and civilization, culture and ancient heritage of a country or region, and or make pilgrimages to religious places based on a religious belief. "When a number of people in a country temporarily leave their place of residence in order to spend the holidays, visit historical monuments, participate in competitions and conferences, visit relatives," said Berkener, a member of the Vienna Institute for Tourism Studies. "And relatives, from one place to another, tourism begins." (Armghan, 2007, 8)

An example of specific and main elements of the city in the Iranian-Islamic period

Mosque: Considering that the Islamic government was formed in the city and its formation was in the mosque for the first time. Therefore, the existence of the Grand

Mosque has become one of the main characteristics of the cities of the Islamic period. And even the first building to be built during the conquest of a city or to replace elements such as a fire temple or a temple was the Grand Mosque, which was considered the biological center of the city and meant the presence of the Islamic State in that area. In addition to its religious role, the Grand Mosque had a very strong political and social role, so that the decree of dismissal and appointment of the agents of the Islamic State was read there, and even the crises of the government were discussed there. However, the Grand Mosque in the city of the Islamic era underwent changes that were in four periods. (Habibi, 2008, p. 42)

The first period; The mosque had a simple and unaffected form and was with simple architecture and materials and was built next to the main square of the city and was modeled on the Prophet's Mosque in Medina.

The second period; During this period, the Islamic state was the Umayyads, and the mosque was built in full relation to the power of the Islamic period, next to the government headquarters, and even near the bazaar and the public street. And even the formation of the first schools next to the Grand Mosque was during this period. (Habibi, 2008, p. 43)

The third period; It was the period of the Abbasid caliphate and it was the beginning of cultural-political rule and a general change took place in the Islamic state and political power was separated from religious power and political power was superior to religious power. Comprehensive means that it is created in the city. That is, where there is no mosque, it is not called despite the large population of the city. (Habibi, from the city to the city, 2008, p. 44)

The fourth period; It was a period of strong local government, and other mosques were built in accordance with local materials and resources, and took on a local and regional color and smell. And the magnificent appearance and memorial architecture play a key role in the construction of the mosque. And the neighborhoods have suffered from ethnic and racial and religious divisions, and each neighborhood has its own mosque, and there was a school next to the mosque. In the Islamic state, religious ideals were separated from political goals and wherever there was no mosque, it was not called a city, and during this period, the mosque served as an ideological-political center.

In general, mosques in Islamic cities have linked religion with government. Illusion has been a divine and religious pole. The mosque in the city is considered as the most important public building, but by replacing the mosque in the urban context, it can be divided into two major categories: Religious, social, economic, and urban factors have changed. The second is the mosques, whose space has been determined and stabilized once and for all with a previous and discrete design of the adjacent physical fabric. Comprehensive mosques, as the most important turning point, had the most attraction; Because it has been in a spatial relationship with the market (Habibi, 2008, p. 45).

Market

The establishment of the Islamic State and its domination of countries with extensive trade laws and regulations, caused this government to benefit from all the laws and regulations on the one hand and to harmonize it with the legal standards on the other hand. And Islam spread from Spain to China's neighborhood, and the existence of Sharia-based trade and commercial rules and regulations, relative security within the empire, made trade more prosperous in the second to seventh centuries. And the juxtaposition of the square and the bazaar in this period have been as two basic elements in organizing the city and even as the main elements of the cities of the Islamic period, like the Sassanid period, the city has been overlooked from the bazaar and the main square of the city.

The seat of government is drawn to the wall and fortification and then beyond it to a reasonable extent. And the bazaar revolves around the school and the mosque, which means that the bazaar also has a hierarchy, including the bazaar of candle sellers and perfumers, the seal and rosary sellers, which is located within the mosque, and then the booksellers, money changers. And it is the leather makers and bookbinders who surround the school or schools, and the school opens its doors in the main rows of these markets, and then there is the bazaar of cloth sellers or Qaisaria. This is due to the importance of the textile industry and its important economic role in the city's commercial center.

Every bazaar has doors and gates, but the gates of Caesarea Bazaar are stronger and after Caesarea Bazaar, there is a market for carpenters, blacksmiths, locksmiths, tin makers and blacksmiths. Along with the fortifications of the city, daily and seasonal markets are expanding. And industries that were dependent on the market and had pollution are at a reasonable distance from the fort, such as pottery, dyeing, etc. The bazaar, like the mosque, started from very simple forms and has evolved over time and has reached full complexity. And it becomes like a local mosque and the bazaar changes over time and creates new bazaars. And the bazaar can be mentioned as the backbone of the city of the Islamic period. And the existence of markets full of goods and people shows the economic power of the cities of the Islamic period. That the Islamic State itself is one of the founders and strengtheners of the market and not because it needs the rotation of goods and capital of its subordinate people. But because it is the original owner of capital and goods. (Habibi, 2008, pp. 46-48)

Markets are one of the most important components of productive-commercial cities and have played a significant role in the growth and development of cities, especially Iranian-Islamic cities. The initial core of most markets was formed near one of the most advanced gates of the city, which in the early market was behind the city gates, which was gradually extended into the city. They were constantly connected by the bazaar. Gradually, more residential neighborhoods were created around the bazaar and the bazaars grew linearly, and one of the roles of the bazaar was the social, religious and political role of the bazaar and the bazaars. In addition, the bazaars had a spirit of cooperation and cooperation in local and urban affairs.

Also, the markets have had a magnificent and unique architecture and have been diverse in Iran in terms of following the territorial, geographical and climatic conditions. For example, in the southern coastal areas, due to climatic conditions and rainfall, there is no brick roof. It is usually covered with a mat and straw. At the same time, it has prevented the direct influence of the sun, so the architecture of the markets has been tailored to the specific climate of each region. Important components of markets should be the main and sub-orders, bazaars, inns and Timchehs. And the most common axes of communication in the traditional urban context has been the market order.

But in the bazaar, there were activities that met the needs of travelers, such as: Palandozan Bazaar, Garmabeh, Caravanserai and the order of the bazaars meet in a place known as Chaharsooq. Chaharsooq is usually an open space with high arches and magnificent architecture and the location of large merchants' rooms or mints. Shiraz Bazaar is one of the most complete, well-equipped and also the most interesting bazaar in Iran in terms of architectural style.

Citadel

One of the main constituent elements that formed the construction of Islamic cities was the citadel. The citadel was a collection of buildings that housed the city's governmental organization and organization, as well as the seat of the ruler or sultan. And once a city became the capital or became important, a citadel was built there. In cities that were administrative-military, the citadel became more important. And the space of the citadel could be created in two places according to the geographical location and the roads of the surrounding area. Due to the lack of such a place near or connected to the city fort, access to a way out of the city to enter and exit the citadel was easily possible. Also, in the cities of Iran, the citadel was in a spatial connection with the bazaar and the Grand Mosque. In addition to the citadel as the seat of government in the city, we must mention the fortifications and protective walls of the city in ancient times. The construction of towers and fortifications and sometimes the construction of ditches to protect cities.

And the element of the citadel, which as a self-sufficient part was usually built on a height and protected by walls. In some cities, such as Tehran, the citadel has elaborate facilities and equipment and also houses elements such as schools, Dar al-Fonun, grain warehouses, government treasuries, royal palaces, palaces and squares. Even the citadel in Tehran as a political center was located near its market. (Mashhadizadeh Dehaghani, 1995, 269)

Caravanserais

Another important element of Islamic cities, which also occupied a large space, was the caravanserai. In terms of function and method of use, caravanserais can be divided into three categories: one is the caravanserai between the cities, the other is the caravanserai outside the city walls and the third Urban caravanserais. The third type that is considered in this article is better and more complete architecture, components and spaces than the other two types. The establishment of caravanserais has gained special importance and prosperity since the Safavid era.

Of course, the caravanserais were a place for the exchange of goods and a place for travelers to stay, and the caravanserais that were in the market were also called Timcheh. Depending on their role and position, the caravanserais also had service elements such as bakeries, baths, stables, fodder warehouses, water tanks, mills, and shops. Of course, this was the characteristic of caravanserais outside the city, which also serve the surrounding villages (Mashhadizadeh Dehaghani, 1995, p. 316).

***House:** It is also a large building complex that is created along a large yard and includes special warehouses and rooms to control and handle a variety of goods. And it belongs to different traders and has a separate way of communication with the market, and through this, people enter the yard with their livestock and unload their load in the yard. (Mashhadizadeh Dehaghani, 1995, p. 270)

*** Team:** It is a building with a closed form and an indoor space without a yard and a large dock, which is often built on two floors. (Mashhadizadeh Dehaghani, 1995, pp. 270)

*** Timcheh:** has a smaller volume than the team and formed a more specific cell of the market, which today still retains its importance and role by transforming into passages or buildings in the style of new architecture. (Mashhadizadeh Dehaghani, 1995, p. 270)

Baths: One of the elements in cities are baths, which were created as health centers in all neighborhoods and existed as a necessity in the culture and thought of the people. They have been one of the most important buildings after the mosque and the school and have been considered as an essential element in the central areas of each part of the city. The establishment of baths in different cities spread with the spread of Islam in Iran. It is even said that baths were one of the main factors in determining the population of each city. So that the residents of several houses take a bath and on average there are several people living in each house. Baths in cities as an urban element occupy space in the neighborhoods of intersections and public places, and each bath serves a part of the city population, baths in terms of architecture and occupy space is significant. (Mashhadizadeh Dehaghani, 1995, 355)

Neighborhoods: Despite the fact that at the beginning of the formation of the Islamic state, creating any social stratification based on ethnicity, race, tradition and . . . It is denied, but when a decentralized Islamic empire is formed, reference to the ancient urban social organization in the conquered lands is made quickly. And there is no news about the formation of a neighborhood according to the system of special shortcomings or citizenship. And we see a combination of three urban, rural and Illyrian communities. And for the first time in the history of urbanization and urbanism in the lands where Islamic-Iranian culture has been prevalent, we encounter neighborhoods whose inhabitants, that is, urban society, is a multi-faceted and intertwined link with the inhabitants of villages of the same religion, they have the same language, tribe and ... that is, they have a rural community. Islam and local government are doubly intense. Conflicts and confrontations of these agents in order to gain more power in the government, in practice, they confront each other in confrontations and contradictions full of violence in urban areas.

The existence of conflicting neighborhoods is a prominent feature of the cities of the Islamic period, regardless of their location. Neighborhoods that each had its own mosque, bazaar, school, water storage, aqueduct, bath, etc., and even in special cases, in addition to a special gate to communicate with the outside of the city, also had a gate to enter the neighborhood and neighborhoods in conflict born and playing a historic role throughout the city's tumultuous history, the neighborhood's connection to its tribal organization was far greater than its connection to neighboring neighborhoods. Such a physical formation that arises from differences and sometimes severe socio-cultural contradictions. It creates a new concept of neighborhood. (Habibi, 2003, pp. 46-49)

Reservoirs: Reservoirs are ponds or indoor pools that are usually built underground to store water. In low-water and desert areas, the reservoir is filled with rainwater or seasonal streams. Water is usually stored in winter and used in summer. Reservoirs are among the facilities related to the aqueduct. How to build a water reservoir, purify and insulate it is in accordance with engineering and scientific principles? Physical and chemical methods are used for purification.

These methods include the deposition of waste materials, the addition of a certain volume of salt to decompose and the disinfection of chlorine released, the use of calcareous compounds for disinfection, and the use of charcoal bags for deodorization. One of the most famous reservoirs is Amir Chaghmaq Reservoir in Yazd and Sardar Bozorg is the largest single-dome reservoir in Iran in Qazvin. Qazvin is one of the cities in Iran that has dozens of reservoirs and many of them still remain today. Qazvin Reservoirs, which are also visited by tourists, include the reservoirs of Sardar Bozorg, Haj Kazem, Sardar Kuchak Reservoir, Molaverdikhani, Zanan Bazaar and the Grand Mosque. (Varjavand, 2008, Wikipedia)

Investigation of the position of the old texture in the spatial organization and physical structure of today's city

The city acts as a living body in its existence. This living body can be described as a living thing, a being whose vital tissues and biological organs are based on its ossification. Therefore, the physical structure of the city can be called the ossification of the city. The ossification of the city is the same network of roads and the flesh and skin are the same neighborhoods and uses of the city that are placed in this system (ossification) according to their importance. Neighborhoods create urban space in their spatial cohesion, neighborhoods are special and independent spaces, each of which has been formed and consolidated according to the importance of their existence.

Today's observations and past maps of this texture indicate an organic and intertwined structure. In this context, the market is in the heart of the city and the shape of the market and the city are affected by each other (the market and other elements of the urban index have been completely influential in the orientation of buildings and the surrounding urban context. Newly established streets, etc.). The bazaar has been developed in the city with rows, houses, taches, etc., and its development has created passages, squares, and dead ends in the city. The bazaar, with all its components, has been an urban element that has met the needs of the city at the macro level and has acted as the main thoroughfare of the city.

It was also a link with other cities and was the heart of the city's economy. What remains of this cohesive system today is the cutting off of neighborhoods by the streets and the change of use of neighborhood centers and the violation of the specific boundaries of each neighborhood. Today, with the creation of streets, commercial activities in the city have become decentralized and the uses that are appropriate for the day and the incompatibility of these uses with the market have disturbed the traditional shape of the market and reduced its power and certainty. The activities are mainly focused on the wall and the edge of the street. A number of old houses with original architecture and built with local materials have been abandoned and are now being demolished.

Recognition of elements, spaces, orders, squares, passages, axes, collections, edges and valuable historical, cultural, physical and religious areas

Urban spaces are the main place of events that connect today with tomorrow. Urban spaces are places of movement between the past, present and future, which include the four basic elements of inhabitants, man-made elements, relationships and time. Elements and components that existed in the old cities of Iran. In their interrelationships, they had an integrated and interconnected spatial relationship. The main elements and pattern of formation of many settlements in Iran in the early stages of their development were: shrine or sacred element, cemetery, gate, market, square, source or path of water, religious buildings, residential units, castles, farms and gardens and roads.

Urban spaces are the open and covered spaces of the communication network, i.e. roads, squares and entrance spaces. Roads include markets and passages. The bazaar has always been the center of the city, and in addition to being the center of economic activity, it has been the center of social, religious, and cultural spaces and activities. He played an important and decisive role in the economic and political destiny of the city. Of course, not all public-city activities are limited to the spaces mentioned. Rather, many of them extended into urban-architectural structures such as courtyards of mosques and other large mosques, tombs, coffee houses, tombs, tombs, as well as natural green spaces around the city. (Ivan Naghsh Jahan Consulting Engineers, 2008, Volume One, Page 87)

One of the important basic features of urban spaces is their dynamism and variability; Because these spaces were the arena of economic and social activities and changed and changed as a result of the changes that took place in the city. In order to identify and introduce the spaces of the city, one must also pay attention to their changes and transformations. Especially in the last half century, due to economic and social changes, many wonderful changes have taken place in many areas. The main elements of the central texture of desert cities include the Grand Mosque, bazaar, seminary, square or lean, bath and caravanserai, and there was a strong spatial connection between these elements. The most important aspects of culturally and historically valuable context in cities are:

1- Existence of a permanent center for unification of the whole body of the city, which guarantees a smooth and healthy urban life.

2- The historical context with its physical, historical and cultural value is the best sign of urban identity.

3- The life and growth of the historical context prevents the deterioration of the city from within as well as its uncontrolled development.

4- Due to its location in the city center, the old part of the city is considered to be the best area for the concentration of commercial, administrative, political and economic services.

The most general methods of preserving and reviving the historical context of the city (attitude to the historical context) with the focus on restoration can be summarized in the following three formats. (Avan Naghsh Jahan Consulting Engineers, 2008, 1, 93)

Examining the general goals of organizing and intervening in the old context

During the restoration, creating a positive view of the old buildings and belonging to it is the most important and fundamental concern. Because merely making physical changes is responsive only in the short term and does not guarantee the survival of the tissue. What leads to positive evaluation and residents' attachment to the tissue alone cannot be due to the aesthetic properties of the tissue. According to Maslow's theory, human physiological needs and comfort are in a lower hierarchy and human emotional and aesthetic needs are in a higher level.

Therefore, in order to create a positive view of old buildings and textures, it is necessary to first meet the comfort needs of texture residents to draw residents' attention to the aesthetic values of texture. (Ivan Naghsh Jahan Consulting Engineers, 2008, 2, 12).

Maslow's theory can also be interpreted as meaning that man has two dimensions, material and spiritual, and to reach the stage of spiritual richness, there is no choice but to pass through the material passage. Satisfying material needs brings a sense of comfort, and meeting spiritual needs creates a sense of comfort. Unless a minimum of human material needs is met, it is not possible to understand and receive spiritual messages, such as beauty.

Therefore, old buildings and textures, which often lack physical elements appropriate to the comfort needs of their occupants compared to new buildings, are not desirable for them, and this causes residents to be oblivious to the beauty of these textures. Visitors to these textures, who have met their material needs from another physical space, are able to receive messages recorded in ancient textures and see them as beautiful. (Ivan Naghsh Jahan Consulting Engineers, 2008, 2, 13)

The old textures at the time of formation, in addition to providing peace of mind, also provided material comfort to its inhabitants. The occupants of these buildings and the provision of these needs is inevitable, and these buildings and structures must compete with new and developing buildings and structures in terms of biological facilities and at least equal. Otherwise, the process of abandonment of old tissues will continue. Maslow also emphasizes this point and addresses the basic human needs.

Figure 22. World with Tourism

According to him, these human needs have a hierarchy according to which they must be met and the physical space must provide it. Perceptual and aesthetic needs the physical environment responds to physiological needs by providing shelter, and by providing material and psychological security, it meets human security needs. The symbolic features and symbolism of the environment respond to the needs of belonging and self-confidence.

Freedom of choice and physical beauty also respond to the perceptual and aesthetic needs of human beings, but it should be noted that the ability to meet the needs at a higher level requires the needs of the lower stages to be met. In practice, it makes it impossible to pay attention and the need to respond at the highest levels, and this has been interpreted as a kind of macro-acculturalization policy in the past. A policy in which the lack of attention to the new physical needs in the old texture, improper constructions without coordination with the adjacent texture, improper and unprincipled repairs of buildings.

The duality between the old and the new tissue and the lack of a favorable physical bond between the two tissues has caused it. In general, if the connection between two buildings in the texture is done slowly and delicately and gradually. Not only will the tissue not have functional problems in terms of traffic and access to services and access, but there will also be no sense of duality in the tissue. But if this connection is weak, the same connection weakness creates mental and functional fragility, which will lead to intensification of duality and separation of two parts of the tissue. (Ivan Naghsh Jahan Consulting Engineers, 2008, 2, 14)

Weaknesses and strategies for reviving old tissue

Due to the long history of urbanization, Iran has numerous ancient cities and ancient settlements in various terrestrial ecosystems. Old structures are on the verge of destruction in many cities and are in dire need of urban restoration planning. The concept of urban restoration has undergone many changes since the 1980s. Urban renovation today introduces the concept of returning home. Which means home, not in the sense of place, city or building, but in the sense of remembering, and living. And return to oneself without the grief of emigration and homelessness.

And return to itself to move towards the future in which the citizen is an active member and will intervene in matters related to the city and himself, which is why this urban restoration is something in which the participation and action of citizens is possible and certain. But it is the revival of actions that unite the lost parts of space and return to the previously known state. That urban restoration tries to create a creative link between the past and the future. And the goal in it is something beyond physical and spatial action. That even this is a scientific-cultural restoration and can be a culture builder. And it tries to transfer the embodied values of the place to the future and the goal is to add contemporary values to the old stable values. And in this transfer of values, what makes sense is the material and spiritual preservation of the hidden values embedded in a lasting historical work. And it can connect the collective memory of the current generation with his historical memory and makes historical memory active in the context of contemporary society and is not presented as a part of history but as an integral part of it. (Yazdani, 2007, 45)

Urban restoration methods

There are five categories of different methods in this field. 1) Protective-sanitary method 2) Protective-decorative method 3) Urban reconstruction method 4) Thematic-local intervention method 5) Comprehensive urban restoration method. The first two methods emphasize the dimension of health and welfare facilities and the aesthetic and visual dimension. And the third method is the opposite of the first two methods, and the fourth method pays attention to the body and activity for limited spaces. In the fifth method, the restoration texture is viewed and studied as a part of a whole, and a coherent connection between the texture and other parts of the city is dealt with, and solutions are formulated. (Yazdani, 2007, p. 47)

Weaknesses: Among the reasons for the failure of the measures taken to revitalize the old tissue are:

1- Lack of proper implementation pattern of tissue regeneration, i.e. there is no suitable method for restoration;
2- Inconsistency of implemented plans with people's expectations;
3- Lack of attention to the prevailing spirit in space in the past and only its physical organization;
4- Lack of definition of a proper life pattern appropriate to the original body;
5- Unorganized tourist facilities and their lack of response to reduce the motivation of tourists has led to domestic and foreign;
6- Continuation of economic-commercial orientation and perspective; It means neglecting sustainable development and lack of cultural-identity view of the

region. And now the main point about the importance of reviving the old texture. Using comprehensive ways for urban restoration that promotes the quantity and quality of cultural and historical heritage through modernization (Yazdani, 2007, 49)

Strategies for regenerative revitalization include

1- Injecting new content into the old body, which means bringing life and life into the space, which is also in line with the spirit of the space.

2- Emphasizing the opportunities in the place: which means using strong religious and cultural capacities. The region as the historical-cultural-religious heart of the city is being revived. In this regard, the capacities and talents of tourism and actual and potential tourism of this part of the historical context can also be helpful.

3- Collective urban memories: which this collective memory makes this historical part as a pillar of urban identity. Which becomes a suitable platform for dialogue between the past and the future.

4- Strengthening the ancient ossification of the complex and its public spaces in relation to the whole city.

5- Emphasis on the historical value and aesthetic value of the texture during revitalization: It is a city to improve the quality and identity of the city.

6- To promote a sense of responsibility towards the collection among the urban people. And the result is that by examining, it can be found that over time, due to the focus of life and construction in the new context, it causes forgetfulness on the old texture, but its revitalization gives identity to the city and in addition to the use of residents It has the potential to attract tourists from all over the country, and this in itself contributes to the economic and social growth and development of the city.

Among them, the existence of valuable historical-cultural elements with identity, the existence of historical-religious memory and the function of identity and memorial of buildings and spaces in this part of the historical area and the existence of historical backbone and center as well as functional values of texture as a religious center. Cultural-tourism (tourism) of the city due to the existence of markets and commercial orders in the context can be considered as strengths in the revitalization and revitalization of this part of the historical context. Finally, the quantitative and qualitative improvement of civic life by revitalization institutionalizes the cultural heritage of the old context as a public heritage and turns it into a cultural heritage. Not except with the participation of citizens in the affairs of the city and related matters. (Yazdani, 2007, p. 52)

Urban tourism and its effects on the appearance of the city and urban spaces

Tourism includes all services and features that are put together to provide what the traveler wants and there are different types, but tourism experts in the world have identified 4 general spaces for it: 1) rural space or green tourism 2) Mountain spaces for mountain sports 3) Coastal space with its own characteristics 4) Urban space is the leader of all these tourists and the most popular of them is urban tourism, which is the basis for tourism development due to its special urban position in many successful countries in this industry. The creation and development of suitable urban spaces, the reconstruction of seemingly abandoned and dead spaces with the intention of reviving the ancient aspects of society are among the effects of urban tourism development. Urban tourists benefit, but urban spaces in contemporary cities are of two categories. (Mirzaei, 2008 Geography and Urban Planning site)

A) Modern and new spaces such as parks, modern sales centers, cultural centers, squares and beaches.

B) Traditional spaces such as bazaars, shrines, cemeteries, gardens, mosques and other old places.

However, from this perspective, the impact of tourism on the body of urban spaces can also be divided as follows;

A) Tourism and the creation and development of modern and attractive urban spaces.

B) Tourism and revitalization of old textures and ancient urban spaces. (Mirzaei, 2008 Geography and Urban Planning Site)

In context (A) it should be said that by creating urban, strong spaces and emphasizing tourism activities and creating more attractions, we can cause the development and connection of different parts of the city. This will be achieved by creating modern accommodation centers, recreational and entertainment spaces, museums and cultural and artistic centers with the aim of developing tourism and benefiting the residents from these services and public satisfaction. As for the case (B) tourism, revival of old textures and revitalization of ancient urban spaces; The old fabric is the center and core of cities. The life of these textures over the past centuries has been the basis of many customs and even existing cultures as well as the urban economic prosperity of the region.

Historic urban spaces have, first of all, the duty to preserve the history, objective and mental identity of the city and to make it current in life. They are the guardians of all memories and memories, and therefore they must include activities in their body that are in line with the mentioned goal. Such spaces succeed in preserving their buildings and historical elements by developing tourism activities and spaces, and make the presence of citizens and tourists' possible day and night.

The presence of people, youth and teenagers, celebrations, all kinds of social demonstrations, performances and the introduction of various cultures by holding ceremonies and celebrations and various forms of social and cultural interactions and everything that reflects the cultural and ethnic diversity of a great nation with thousands of years of history. All these effects that are effective in the growth and promotion of culture and knowledge of a city and nation. In such spaces, it paves the way for the growth of the tourism industry and attract domestic and foreign tourists. Hence, the economy of the old city center is formed with industry and tourism activities and finds new life. The economic strength of the old centers is another issue that is considered as a stimulant and booster by experts.

Sustainable economy with emphasis on tourism in old centers can be a source of economic exploitation of these centers, which is an excuse for the revitalization of old centers and cause the transfer of luxury and superior functions due to the capacity of these spaces to them. In contemporary times, by recognizing the main factors of the destruction of old valuable textures and the compatibility of modern words of architectural design with the ancient language of indigenous architecture, we can play a role in improving the urban environment and its future development directions. The philosophy of revitalizing and repairing damaged and worn-out organs of cities is rooted in this position, and it seems that one of the most effective strategies in revitalizing old tissues is to re-create activities in these areas according to their physical abilities. Creating tourism hubs can be used as one of the methods to revive old urban spaces and attract tourists.

Tourism hubs, if they have the mentioned values and characteristics, can change their surrounding texture. For example, creating commercial activities in the tourist axis in a way that attracts tourists transforms the neighborhood economy and is also effective in changing the use of valuable buildings as places of rest such as hotels and restaurants. From Dr. Mohsen Habibi's point of view, designing tourist sidewalks with the aim of protecting the texture and boosting the tourism industry is one of the most important measures to revitalize this texture. Because the positive results of this action are understood by the residents and exploiters of these areas, it can motivate their conscious intervention and participation to improve and correct the texture. In his view, the importance and value of these ancient textures is such that they have now been able to attract many tourists with limited essential facilities.

Therefore, organizing and rehabilitating these urban centers using sidewalks, in addition to achieving the protective purpose of these structures due to the quality improvement of the relevant spaces and their proper equipment, not only meets the needs of residents for desirable urban spaces for current life but also tourism development wheels. It activates in this context, and this not only increases the social and cultural identity of the context, but also effective and experienced solutions to create and reproduce public spaces and lost identities and dangers removed from the historical contexts of cities. It is Iran.

However, in the field of urban tourism, municipalities have responsibilities that in the process of approving laws and establishing primary organizations related to tourism and tourism, show that their activities are initially in the Ministry of Interior and a large share of their capital and assets are given to municipalities. However, with the formation of independent organizations related to Iran Tourism, the role of municipalities has been formally and legally reduced in this regard, while their duties have increased in practice. In the Municipal Law (approved in 1955 and subsequent amendments), the general duties of municipalities are defined in Article 55 in paragraph 28 or axis, which can be divided into four groups: civil, service, supervisory and social welfare. Among them, service, leisure, recreational and cultural duties have been proposed due to low conditions. (Jahandoost, 2008, p. 84)

Guaranteeing the sustainable development of urban tourism with the protection of old-historical textures

With the advent of modernism and the influx of non-indigenous cultures and the consumption pattern of the advanced world into the country, the new urban culture has been contrasted with the traditional culture and the culture of modernity and modernity has replaced the original and traditional culture, which has led to an identity crisis in cities. And especially among the inhabitants of the old textures of the ancient cities of Iran.

Among the countries with ancient history and civilization, without exaggeration, the cities of Iran may be the only regions that go through such a bitter and unfortunate experience, and a brief look at the historical cities of the world such as Venice, Athens, London, Paris, etc. It confirms the preservation of ancient-historical textures and emphasizes the use of past architectural elements in developed areas. The old buildings and textures represent the culture, art and architecture of the time of their construction, in other words, the historical and cultural form of each country and nation, and in the old memories of nations, they mark the artistic revival of previous civilizations. These buildings are undoubtedly in close connection with the urban context around them at the time of construction and re-examine their full meaning with these civilizational phenomena. It will be a curtain of ambiguity on the old book of history and civilization for the future. (Mo'men Lou, 2009).

Due to the ancient cultural and civilizational background of urban culture, Iran has always witnessed the emergence of many cities in all parts of different areas of its civilization. The relics that have now appeared in the heart of the old textures and have been the beating heart of the great Iranian civilization. Isfahan, Shiraz, Yazd, Tehran, Hamedan, Kerman, Bam, etc. each have masterpieces from different periods of urbanization and urban planning, monuments that have preserved behind millennia and different periods, but Unfortunately, they are now exposed to collapse and public neglect.

The involvement of responsible and irresponsible institutions in the field of old textures has now revealed serious and important concerns in order to protect these national treasures, and if no solution is found for this concern now, surely tomorrow will be nothing but regret. It will not be our income and the ancient culture of Iran. For example, the old texture of Yazd with an area of about 800 hectares can be considered the healthiest, most extensive and most principled texture in Iran, which has experienced different approaches in the field of intervention of those in charge of road networks, but in the meantime ignoring the value The physical, historical, and identity of some contexts, along with the delay in defining the old boundaries and the formulation of new rules and policies in how to intervene in them, have plunged these valuable elements into isolation more and more.

Tehran may not have much history in the field of cultural heritage and ancient context, but its importance is due to the fact that Tehran, as the capital of our country in contemporary history, has been an important base for political and social developments. "Old Tehran, according to the available evidence, had five neighborhoods, each with its own hangouts." has predicted migration.

However, with the election of Tehran as the capital of Iran, the earth has grown so much over the past 200 years that it is considered one of the most polluted and largest capitals in the world, and of course it is necessary to mention that along with urban development, Old textures over time and historical buildings, now in the international arena, these textures have become one of the most valuable urban areas and attract thousands of tourists. Experiences that if the work of planners in the field of urban management and cultural heritage and tourism organizations, will be enjoyed in the not too distant future with the expansion of the tourism industry in our historic cities. In other words, the development of urban and sustainable tourism should be considered as a guarantor of the preservation and protection of old textures, because now the service to old textures has become one of the main concerns of urban management.

From the point of view of Dr. Seyed Mohsen Habibi, designing tourist sidewalks with the aim of protecting the texture and boosting the tourism industry is one of the most important measures to revitalize these textures. Because the positive results of this action are understood by the residents and exploiters of these areas, it can motivate their conscious intervention and participation to improve and correct the texture.

In his view, the importance and value of these ancient textures is such that now, with the most limited necessary facilities, they have been able to attract many tourists. Therefore, organizing and rehabilitating these urban centers using sidewalks, in addition to achieving the protective purpose of these structures due to the quality improvement of the relevant spaces and their proper equipment, not only meets the needs of residents for optimal urban spaces for current life, but also wheels It activates the development of tourism in this context, and this not only increases the social and cultural identity of the context, but also effective and experienced solutions to create and reproduce public spaces and lost identities and memories erased from contexts. Are the old cities of Iran? (Mo'men Lou, 2009).

Perhaps the most important principle in preserving and improving the old texture, especially cities such as Tehran, Isfahan and Shiraz, is to determine its boundaries as a special urban area, because in this case, special criteria can be developed for major controls to preserve the identity of old textures. Determine the functional grading of the passages of the old textures, apply possible widening's in accordance with the old passages, prevent the formation of impenetrable and dense edges in the body of the passages, apply special rules to prevent visual disturbance in the texture and the main body. And the skyline, the identification, preservation and protection of a set of elements that somehow have historical and cultural values and help maintain the spiritual identity of the texture should be considered one of the most important principles that observe the preservation of old textures as material assets. And the spirituality of cities helps. Creating tourism hubs in areas such as bazaars can be used as one of the methods to revive old urban spaces and attract tourists.

Tourism axes, if they have the mentioned values and characteristics, can change their surrounding texture. For example, creating commercial activities in tourist axes in a way that attracts tourists, transforms the neighborhood economy and by changing the use. Valuable buildings such as rest areas such as hotels and restaurants are also effective in this regard. Inspired by expert views and global experiences, Tehran Municipality has taken great steps to protect the old textures and cultural buildings. The plan to introduce old buildings by the Azadi Tower Beautification or washing organization in cooperation with the Cultural Heritage Organization are good examples of this national effort and sustainable interaction with the country's cultural heritage. The large-scale implementation plan of Tehran Bazaar is a seal of approval on the performance of the municipality in the face of the protection of the old texture and shows the potential of urban management in this area.

Even about Tehran Bazaar, Dr. Qalibaf, the mayor of Tehran, said in the Tehran implementation project: "All countries in the world have the experience that the old centers of the city, and in fact the old citadel and the main center of the city, are preserved as cultural heritage. Now that people live and trade there, they spend their best hours of leisure and relaxation emphasizing the preservation of the old city center as the best and most beautiful city. In September 2009, the Tehran City Council passed a resolution to protect the old houses of Tehran. The decree provided for the ownership of old houses to benefit from a set of facilities. Also, according to article 166 of the Fourth Municipal Development Plan, to help the cultural heritage and tourism organization, the mayor's structure's the old texture to protect create the cultural heritage of cities.

The collection of archeological and historical monuments within cities not only needs to be restored and protected in order to preserve the national values remembered by the ancestors of any society, but also by their objective presentation to tourists, cultural tourism is spread and expanded. Although such monuments and buildings will sometimes not regain their former uses, but by presenting cultural-traditional users and by looking at aspects of their tourism, the economic aspects of cultural monuments can be used while preserving their values. Given the importance of the role of the municipality as the custodian of urban management and the cultural heritage organization as the custodian and policy-maker of the preservation and restoration of historical monuments, it seems that the formation of a joint committee to follow the issues of the old context is a necessity today.

The old textures are the center and core of the old cities of Iran and the life of these textures over the past centuries has been the basis of many customs and even the existing culture and economic prosperity of the city and region. Historic urban spaces have, first of all, the duty to preserve the history, objective and mental identity of the city and to make it current in life. They are the guardians of all memories and memorials and therefore they must include activities in their body that are in line with the mentioned goal.

Old tissue; A dictionary to be learned or words to be thrown away; man, since he came into being, has made every effort to meet his needs and overcome natural and social obstacles, and so far some have succeeded. Earned gifts; The result of this effort has mainly led to the emergence of culture and civilization. Culture has been divided into two types of spiritual and material cultures. Material culture; Inventions and discoveries include technical and material means that have survived from the past, and have now reached the present generation, and, with further completion and production, are passed on to future generations. Spiritual culture is a collection of artistic, scientific, ethical, and traditional achievements through a language, script, or a set of different urban-architectural units that are gradually assembled in human settlements, and a set of values. It includes the ideologies and civilizations that have come down to us from the past, and are passed on to the future.

Thus, two main axes in the emergence of culture are confirmed; One is man and the other is the environment around him, which is his place of residence. In this way, man, by interacting with his surroundings, has tried to make the most of the least utilization by using the least facilities, and for this reason, the way of life in a place with a hot and dry climate is different from the way of life is to live in another place, with a hot and humid climate, or cold and humid.

All these factors have led to the emergence of different perspectives in social and urban spaces, including how construction of housing and places used, type of cover, customs and social behaviors. Therefore, it can be inferred that living for a long time in a specific geographical place causes the presentation, hypothesis and production of knowledge that this knowledge sometimes takes place in the mind of society and is passed on to future generations like an identity card and cultural background. The body of historical works is located that the preservation of those works can prevent the anonymity of society.

Urban texture and elements of the old texture

"In general, each city or geographical location, according to its culture, has special buildings and characteristics that everyone, especially officials, should know that an old building, before it represents the technical issue of construction, includes a series of "It is the ways of thinking and acting that tell the way of life of the people of their time." In the old cities of Iran, we can distinguish three spaces with different characteristics from each other: (Dehghannejad, 2010)

1- Private space; Such as the yard and the elements that contain it.

2- Semi-private space - semi-public space in the form of a private dead end, or a porch that leads to several houses.

3- He mentioned public spaces such as mosques, bazaars, squares and public baths. (Dehghannejad, 2010)

The most important private space of the old texture is the house, the features of which have played a role beyond the shelter for its inhabitants. In the past, Iranian houses were not only a refuge for the family; A place for family and local gatherings, a place for education, cooperation and collaboration, group work, use of private green space and healthy recreation, use of the garden and production of some products for personal consumption, and compliments and offerings. It was also common among neighbors. The interior of the house, depending on the ability of the owner of the house, had special decorations and facilities. The manifestation of creativity, efficient architecture and art can be seen in interiors as stated:

"Of the 200 houses studied in Isfahan, two are not exactly alike, and this creativity was either created out of necessity, or using the experiences of predecessors." (Dehghannejad, 2010). The interiors of houses of different floors were different; But we tried to make the exterior of the house (with the exception of some houses that had different entrances) the rest, not different from other houses around. There were side platforms at the entrance, and on the right, a metal ring for women, and on the left; A metal percussion was considered for men, which even its position was not without reason; And to take care of women who want to be accountable behind the door. "To enter different spaces, including the party space and the interior, they had to go through winding corridors called corridors."

All these measures have been taken to limit the invasion and view of strangers into the sanctuary of the family, and their Muharram world. He closed it and when he returned, he opened it again, which showed the attention to privacy at that time, and also used shards of glass on the walls of the houses to prevent thieves from entering the houses. By examining the cultural features present in some of the public elements belonging to historical cities, one can understand the depth of its effect on the human soul and body. This feature can be seen even in the smallest detail of the city, the Saqakhans, which were accompanied by phrases such as "Ya Hussein Shahid" and images of the Karbala incident, whose designer's goal is not only to irrigate the passing body; But also thinks of his soul. (Dehghannejad, 2010)

"Mosque" is one of the important elements of the city that is considered as the most important public building, and in the ideas related to Islamic urban planning, plays the role of an underlying factor. "Mosques, in addition to their religious, social and political roles, also played an educational role before the establishment of independent schools.

(Dehghannejad, 2010). In mosques that have also been educational centers; "It has been taught from elementary education to the most complex philosophical issues." In addition to the practical aspects of building a mosque, special attention was paid to cultural and religious matters, which all people considered themselves obliged to observe. For example:

The worshiper or anyone who wanted to enter the mosque must be clean, and the religious people have adhered to this ritual to the extent that they did not attach any building to the mosque. Around the mosque, there were always passages, bazaars and alleys, so that no one could linger next to it. Also, if a school was to be built next to a mosque, they would constantly build a walkway instead of a room to the fortifications and walls of the mosque; Like the Isfahan Grand Mosque. (Dehghannejad, 2010)

From ancient times, outside the shrine, they provided a place for washing, so that the worshiper could get rid of the filth before entering the shrine, and be ready by washing his body and head and wearing clean clothes. When the preparations were made, and the worshiper set foot in the mosque, he was immersed in the grandeur and beauty and art that can be seen on the walls, domes, minarets, pulpits, altars and inscriptions around him. All this caused the person, under the influence of the spiritual atmosphere that prevailed, to feel pleasure, and to hesitate to leave the mosque, and to try to discover the secrets of the building by reading the building lines. For example, in the Isfahan Grand Mosque, he found that he had written a poem by putting bricks together. (Dehghannejad, 2010)

The market as a public element, in addition to its economic role, has been a place of gathering, a place of trade, a hotbed of riots and strikes, a policy-making place, a place of information, communication and acquaintance, and sometimes a role of a place of security has also played. (Dehghannejad, 2010)

In Iran, the markets of the old and historic cities, according to Carlaserna, are the Italian tourist: a public meeting place, where they trade, talk about private or commercial matters, and discuss public or governmental affairs. comment; In a word, in his opinion, the market is both the center of transactions and the council. (Dehghannejad, 2010)

Based on the architectural features of the market and its location; In addition to meeting the needs of people in cities, the spiritual, human and moral aspects of society were also considered, so that it can be said that the indoor market of Iranian cities with different climates is still the most suitable place for Iranian craftsmen and buyers. Healthy competition among vendors, the power of choice for the customer, especially in the markets, strengthening the spiritual and Islamic dimension and joint efforts and altruism through proximity to mosques, especially the Grand Mosque, ensuring the comfort and security of people, especially women, because of feelings The responsibility of the sellers towards the buyer, the feeling of love and affection in the drawing of brotherhood between the seller and the buyer, which is probably all because of the market atmosphere, it offered values to the people, and those values still exist in the old markets. Another common element of the old texture was the bath. The existence of many baths in different places shows the importance of Iranians for cleanliness, which, of course, has become more evident after Islam. "The function of the bath is not limited to cleaning the body.

The bathhouse is a gathering place for different social strata. That the entrance and exit of men and women were separate and there were places for their better interaction, which is why people, especially women, talked for hours in the bathroom eagerly. "And they performed rituals such as hanabandan, bridal bath, delivery bath, etc." Thus in the past, the bath has paid attention to both the physical dimension of human beings and the human soul, so that the baths had a space that a person did not quickly reach the outside environment when coming from the bath, but this space with It took a while, and this prevented the person from getting sick, which showed that he was paying attention to people's health. (Dehghannejad, 2010, Raskhoon Cultural-Artistic Site)

In the construction of the bath, religious issues are not overlooked, and "because of the lack of use of angry water, even if the bath was next to the aqueduct, a well was dug to use its water." (Dehghannejad, 2010)

The square is also one of the general elements of old cities, for example, Naghsh Jahan Square in Isfahan is considered as a square that "has played the role of gathering houses or urban and neighborhood elements. This square, physically, is part of the spatial structure of the city, and socially, this complex has a place in the souls of citizens, and has been a support for Iranian citizens. (Dehghannejad, 2010).

According to some urban planning experts, Naghsh Jahan Square, which is 520 meters long and 152 meters wide, i.e. one and a half kilometers of the body, is broken only in four points; It shows the ultimate humility and restraint of the architect of this field in the face of unprincipled emotions, and it is in the shadow of these principles that architecture is formed, and inherited, the thinking of the predecessors becomes the next intellectual capital, and This is not possible except with a continuous, continuous flow and with codified principles. The relationship between the components and elements of the city, both material and immaterial, which are changed by changing one, one or more other elements, should also be considered to avoid inappropriate interference in the old context. (Dehghannejad, 2010)

Now, given that the old context is usually a serious obstacle to change for municipalities; But from an academic point of view, it is a research book; That is, just as designing a question and delivering it to the answer is what happens in a book, so it is in the collection of historical monuments. This feature shows and clarifies the process he went through in order to arrive at a question from an answer in his time. That is, every work of art has something to say, words that are intact and reliable; But the remarkable thing is that only the sighted have the power to read this book. Not every stranger and stranger will find their way to this collection. (Dehghannejad, 2010)

How the old texture changes

Involvement in the old fabric increased since the industrial revolution and the entry of cars into urban textures and urban population growth, affecting almost every country in the world, except that developed countries did not have increasing population growth, and They also designed cars that were more compatible with their older textures. In the meantime, all the cities of Iran more or less underwent changes and transformations; But these developments have been many times more in immigrant and industrial cities such as Isfahan. As one of the officials of the Cultural Heritage and Tourism Organization, in his description of Isfahan, says: "Until a few decades ago,

Isfahan was a large historical-cultural museum in which humans also lived; "But today it has become an industrial city that, ironically, also houses a number of historical monuments." (Dehghannejad, 2010)

The destruction or evolution of the old texture is faced with different theories. Some believe that the old texture is composed of two worn and valuable parts, and as much as they emphasize the preservation of the valuable part, they insist on the destruction of the worn part. Others contain the old context of historical and cultural background. They know that destroying it means destroying that history, and ultimately anonymity. (Dehghannejad, 2010). But some believe that the destruction of works, if the mind of the community is awake, causes complete anonymity.

It cannot be, and its memories are always alive; Because the past never dies; Rather, it is a constant part of existence. This group of experts seems to pay attention to the key role of history, and say: history should play an important role, and reveal the forgotten existential factors of our cities; Because basically one of the tasks of history is to help us live in a wider space of insights and senses. But it should be noted that one of the necessary tools for history and historians to fulfill this task, is the existence of these historical and ancient buildings. (Dehghannejad, 2010)

This disagreement has also permeated relevant government agencies, such as the municipality and the Cultural Heritage and Tourism Organization. They compete with each other to bring their beliefs to fruition; In Isfahan, in order to widen Hakim Street, they attempted to demolish Khosrow Agha Bath, and after protests from other organizations such as cultural heritage and antiquities lovers, the decision was made to rebuild this destroyed building, based on the remaining ruins, and in accordance with they got the existing designs!

Of course, it is not logical to reject the change in the old tissues; But we must emphasize the correct connection between tradition and modernity and know that this building of the Safavid period, which is considered old in our time, was new in the time of Shah Abbas, and must be next to the historical buildings of that time. Which was built during the Seljuk period, including the Grand Mosque; But an attempt has been made to establish a logical and principled connection between these two old and new contexts;

Not that, like the present age, by building a world-view tower, in the sense of "putting a thorn in the side of the role of the world." (Dehghannejad, 2010)

It seems that in order for any intervention in the old context to be subject to human principles and laws, we must seek the system of this context, and we must have a history of living in this context. Keep in touch, and make contact with the human dignity of that context, and finally give a human process to our executive work. (Quoted from the theme: Mojabi, 2008, p. 143)

"In creating the connection between tradition and modernity in architecture and urban planning, the effect of emotion on human life should not be neglected; An effect that, although it may not be easy to determine at times; But especially in shaping our current culture, it has great importance and meaning "(Giden, 1986, p. 679)

Attention to culture and traditions has been considered in the urban planning of the world, especially with the emergence of postmodernist thinking, most of the believers in this school express their reluctance to new methods of urban planning. "Some of today's committed architects in India are trying to transform contemporary architecture in a way that has both cultural and historical content and meets the needs of the 21st century." Raj Rawal wrote in 1985: "The effort of our generation is to create a connection between the architecture of the past and the contemporary era, while maintaining its chain" (Rawal, 2000, pp. 58-61)

But we conclude from this discussion; The elements and components of historic cities and ancient contexts contain information that reveals the cultural role of these elements in relation to other roles, and indicates that our ancestors did not build without thinking, and for Any architecture; There has been economic, social, cultural and even political justification. In the meantime, the role of the people and the beliefs and cultural aspirations have been very key, and while using the experiences of their predecessors, they have not been far from creativity and initiative.

The connection between urban components and elements in past constructions, it has received the most attention, and any development, without considering other components, has not been considered. Even today, in order to preserve culture, one must study traditional architecture, and recognize its elements and factors, and deal with it through contemporary language and expression. The presence of historical elements in the old context has benefits beyond the transmission of culture, which will be achieved by discussing and studying students and researchers in this field, and encouraging people to participate in the preservation of these ancient-historical buildings.

Figure 23. Mayan Ruins in Mexico

In general, the old texture of the historical cities of Iran are able to depict the artistic and historical features of the country, and to display the customs, traditions and culture of the society in the form of old-historical buildings, components and elements; But this context today also faces many problems that can be generally addressed to welfare-development problems, such as lack of access; Lack of urban facilities and social problems, such as old-fashioned immigration; He mentioned addiction, delinquency, and so on. Some experts and city officials have seen the way to deal with these problems in the changes of the old texture, and some because of the cultural-historical importance of this part of the city, oppose any change. This has led to fundamental differences between different organizations in the city in this regard. (Dehghannejad, 2010)

WO model

In the field of urban tourism management analysis, different models can be used. Among the quality models, the most efficient model is the SWOT model. This model is in fact the abbreviation of four analytical factors in this model.

Which are strengths, weaknesses, opportunities, threats (threuts). In fact, the analysis of strengths and weakness in the internal environment and the analysis of opportunities and threats from the external environment is a systematic process that provides support for the decision-making situation, forming a qualitative model (SWOT) (wheelen.1995.341).

This model can be an initial stage of an analysis with the ultimate goal of presenting and adopting the necessary policies to fit between internal and external factors (kajanus.2000.718). When SWOT is fully used, it can provide a good basis for policy formulation. (mcdonalo.1993.143). Strengths, Weaknesses, and Opportunities-Threats Analysis provides the formation of preliminary goals to the development strategy and a preliminary ranking of actions that facilitate the achievement of short-term, medium-term and long-term goals. After the internal study, this model evaluates the opportunities and threats of the external environment and the external study identifies issues such as economic, political, cultural conditions ... The art of strategic management is to be able to get the best combination, which is the result of these studies, for planning.

In the second case, when the organization has weaknesses in the environment and is ready for operation (there are opportunities for the organization in the environment), management must use a favorable environment by changing the type of service or product and overcome the weaknesses of its organization. For example, an organization that is financially weak, but there are customers in that environment who overcome financial problems by subscribing customers and receiving money in advance. (Kaviani, 2010)

In the third case, when the organization has sufficient strengths, but the environment is not favorable, arrangements should be made so that environmental threats become opportunities and do not cause harm to the organization. Supposedly, an organization is in the third position due to the high price of its services or goods and the environment in which customer revenue has decreased. This organization can solve this problem by selling in installments and create opportunities for itself from an unfavorable environment, because those who have a low income will welcome this type of sale a lot. (Kaviani, 2010). In the fourth case, when the organization is weak and the environment is threatening, the organization should abandon its current activities and seek to continue working in another market with different products. External in the SWOT model, the strategies of the organization are realistically adjusted according to the internal facilities and external opportunities and the fourfold future and the determination of weaknesses, strengths, opportunities and threats is not done only with the comments of officials and managers, but with Obtaining accurate and documented information and statistics is defined without any special orientation and prejudice of these situations.

Figure 24. President Andrés Manuel López Obrador has a dream for the Yucatan Peninsula. He wants to build a train that will leverage the tourism economy of Cancun by bringing more visitors inland to the colonial cities, Mayan villages and archaeological sites that dot the region

References

Armaghan, Simin. 2007. The book Tourism and its Role in Geography, Islamic Azad University, Islamshahr Branch, Faculty Member, Islamic Azad University, Islamshahr Branch.

Alwani, Seyed Mehdi. Dahti, Zohreh 1994. Principles and Foundations of Tourism, Publications of the Foundation of the Oppressed and Veterans of the Islamic Revolution, Tehran.

Meteorological Department of Semnan Province, 2007. Climatic Zoning of the Province, Volume 3.

Eftekhari and Mahdavi, 2006, Intervention in old and dilapidated urban contexts (history and backgrounds) Special Letter No. 14 of the Municipalities of the Ministry of Interior, Municipalities Organization.

Bakhtiari, Saeed. 2002, Atlas of Iranian Roads (Institute of Geography and Cartography of Geology).

Bani Asadi, Ali, 1993, Semnan Province TV, Iran Statistics Center Publications, Tehran.

Papli Yazdi, Mohammad Hussein. Saghaei, Mehdi 2006. Tourism (nature and concepts). Publications of the Ministry of Culture and Islamic Guidance.

Taqvaee, Massoud, Ahmadi, Abdul Hussein. 2003, Determining and analyzing the levels of enjoyment of cities, villages and districts of Kermanshah province, Quarterly Journal of Rural and Development, Publications of the Research Center for Rural Issues. The sixth year. Nos. 1 and 2.

Tavassoli, Mahmoud, 2008, Quarterly Journal of Urban Development and Improvement, Haft Shahr, Ministry of Housing and Urban Development No. 2.

Jedari Eyvazi, Jamshid. 1994. Basics of Water Geography, Payame Noor University Press.

Jedari Eyvazi, Jamshid. 2004. Geomophology of Iran, Payame Noor University Press.

Javan, Jafar. Summer 2000. Quarterly Journal of Population and Development. Publications of the Civil Registration Organization. No. 32

Jahandoost, Rasoul (Master of Geography, University of Tehran), Koohstani, Ghasem (Master of Civil Engineering, Khajeh Nasir al-Din Tusi University), 2008, city tourism and its effects on television and urban space, geography site.

Hafiz Nia, Mohammad Reza. 2004. Introduction to research method in humanities, Samat Publications.

Habibi, Seyed Mohsen. 2008. The book from the city to the city (a historical analysis of the concept of the city and its physical appearance and thinking and impact), University of Tehran Press.

Khodadadi, Raheleh, 2009, Research project on population change in Semnan city, Consultant: Yousef Ali Ziari, Young Researchers Club of Islamic Azad University, Semnan Branch.

Khairuddin, Ali, February 17, 2007. National Conference on Capabilities, Obstacles and Strategies for Tourism Development in Semnan Province, in collaboration with the Cultural Heritage Organization, Semnan University.

Damghanian, Hussein. February 17th, 2007, National Conference on Capabilities and Strategies for Tourism Development in Semnan Province, in collaboration with the Cultural Heritage Organization. Semnan University.

Dehghan Nejad, Fatemeh. 2010. The old context of the dictionary to be learned or words to be messed up, discarded. Raskhoon Cultural and Artistic Institute website.

Rasooli, Muhammad. 1997, Proceedings of the 14th Specialized Conference on Urban Textures and Improvement and Renovation of Old Textures. Department of Housing and Urban Development.

Rezvani, Ali Asghar. 2002, Geography and Tourism Industry, Payame Noor Publications.

Romina, Nadia. 2005. Master Thesis; A Study of Land Use Developments in the Old Texture (Case Study of Semnan), Supervisor; Yousef Ali Ziari, Consultant; Zeinab Karkabadi.

Iran Independent Morning Newspaper, July 4, 2007, Afarinesh Sanandaj, Afarinesh Reporter.

Rawal, Raj, Winter 2000. Architecture rises from the heart of society and people, Abadi Magazine. Seventh year. Nos. 27 and 28, special building festival.

Rahnamaei, Mohammad Taqi. Year 1990. Polycopy. Geography of leisure and tourism, geography of Tehran University.

Rahnamaei, Mohammad Taghi, 1995, Development of the old urban context (concepts, history and strategies), Proceedings of the Second Housing Seminar, Ministry of Housing and Urban Development.

Meteorological Organization of Iran, 2008 and General Meteorological Department of Semnan Province.

Dezful Cultural Heritage Organization, 2010, Dezful Shenasi site, avaei.ir site

Statistical Yearbook of Semnan Province, 2008, Deputy of Planning, Program and Budget of Semnan Province. (Written by Abolfazl Semnani)

Statistical yearbook of Semnan province, 1966 to 2008, Management and Planning Organization of Semnan province.

Saghaei, Mehdi. October 12, 2009, Tourism Industry, Tourism Studies Site.

Semnan Province TV, 2002, Volume 2, Office of the Deputy and Planning of Semnan Province.

Housing and Urban Development Organization. Year 2008 (Maps)

Shafaghi, Ali, 1997, Strategies for Tourism Development in the Islamic Republic of Iran, Proceedings of the First Tourism Conference in the Islamic Republic of Iran. Volume One. Kish.

Shokouei, Hussein. 1975. Introduction to Tourism Geography, Social Research Publications, Tehran.

Shokouei, Hussein. 1998. New Thoughts in Geography. Tehran Geological Survey.

Sadeghi, Hamid Reza. 2009, Mayor of Niasar, Kashan, Niasar book in the old garden of Iran. Mirdashti Publications.

Sayadjoo, Ali, 1989-88, Master Thesis; The Economic Role of Cooperatives in Sustainable Rural Development (Case Study of Semnan County), Supervisor; Abbas Arghan, Consultant; Abbas Bakhshandeh Nosrat, Islamic Azad University, Semnan Branch.

Zargham, Hamid, 1997, Strategies for Tourism Development in the Islamic Republic of Iran, Proceedings of the First Tourism Conference in the Islamic Republic of Iran. Volume One. Kish.

Tasian, Mohammad Javad, 2004, Roundtable on the crisis of old and worn urban structures, special letter No. 14 of the municipality, Ministry of Interior, Municipalities Organization.

Comprehensive tourism plan of Semnan province. 2006. Cultural Heritage and Tourism Organization.

Master plan of Semnan city. 1995. General Department of Housing and Urban Development of Semnan Province. Volume II.

Civil Development and Influence Development Plan (Semnan City Comprehensive), 2009, Architect and Urban Planning Consulting Engineers.

Farzin, Mohammad Reza. 2004 and 2008. Tourism Economics, Commercial Publishing Company affiliated to the Institute of Business Studies and Research.

Farzam, Farhad, 1997, Proceedings of the Fourth Specialized Conference on Urban Textures, Ministry of Housing and Urban Development.

Flamaki, Mohammad Mansour, 2001 Revitalization of ancient urban buildings and spaces. University of Tehran Press.

Kazemi, Mehdi, 2006, Tourism Management, Ministry of Culture and Islamic Guidance Publications.

Kaviani. November 19, 2010, MBA management site and SWOT method analysis in strategic planning and SWOT model.

Kiden, Zickfried, 1986, Space; Time and Architecture. Translated by Manouchehr Mozini, Tehran, third edition.

Goli, Ali, Saghaei, Mehdi, Ezatollah Humafi, September 7, 2009, Application of MS_SWOT model in tourism management, Case study of Mashhad metropolis, Journal of Geography and Development, No. 14.

Goli, Ali. September 21, 2009, Tourism and the place of tourism in Iran, it, the site of tourism and tourism.

Statistics Center of Iran. Year 2007. Principles and foundations of geography of Tehran population. Qom Publication. The tenth edition.

Statistics Center of Iran, 2006, Detailed Results of the General Census of Population and Housing. Deputy of Planning and Governor of Semnan.

Mashhadizadeh Dehaghani, Nasser. 1995. Analysis of urban planning features in Iran. Iran University of Science and Technology Publications. Bahman Print.

Mojabi, Seyed Mehdi. 2008, Social man and the old context, economic man and the metropolis, abstracts of the seminar on the continuation of life in the old context of Iranian cities. Faculty of Architecture and Urban Planning, Department of Restoration and Restoration of Buildings and Textures

Mukhlesi, Mohammad Ali. New edition of 2009, book of historical monuments of Semnan.

Armanshahr Consulting Engineers, 2009, Semnan County Master Plan, Employer of Semnan County Housing and Urban Development Organization.

Part Consulting Engineering, 1991, Semnan Governor's Technical Office. Master plan of Semnan city, Shahroud.

Avan Naghsh Jahan Consulting Engineer, 2009, Managing Director Nasser Mashhadizadeh, Semnan worn-out tissue management plan, employer; Cultural Heritage Organization. Volumes one and two.

Talesh Consulting Engineers. 1994. How to intervene and deal with historical monuments and urban centers of the world.

Monshizadeh, Rahmat A. .... 1997, Principles and Foundations of Tourism, Samat Publications.

Mansouri, Seyed Amir. 2004, Master Thesis; Investigating and recognizing the effects of Semnan city development on the environment of Islamic Azad University, Semnan Branch. Supervisor: Yousef Ali Ziari, Consultant: Zeinab Karkabadi.

Mo'men Lu, Ali. February 17th, 2009, Guarantee and sustainable development of urban tourism with the protection of historical contexts

Mirzaei, Hossein. March 12, 2008, site of geography and urban planning, tourism and revitalization of historical texture.

Nobakht, Mohammad Baqir. Victorious, inspiration. 2008, Development of Tourism Industry in Iran; Obstacles and Solutions, Islamic Azad University Vice Chancellor for Publications.

Nouri Nasab, Jafar. 2008, Master Thesis: Changing the cultivation pattern in the village of North Khorramarud, Supervisor: Abbas Bakhshandeh Nosrat, Consultant: Abbas Arghan.

Varjavand, Parviz, 2008, Reservoirs in Iran and different parts of reservoirs, Wikipedia.

Honarparvar, Emad. May 26, 2010. Tourism ¬ Collection in Iran Articles, reports and tourism news. www.business_iran.persianblog.ir

Yazdani, Maeda. 2007, Master Thesis, Architecture, Supervisor, Seyed Amir Mansouri, Weaknesses of Historical Texture Restoration Strategies, Iran University of Science and Technology.

Ebrahimzadeh, A., Ebadi Jokandan, 2008, An analysis of the spatial distribution of green space use in the three urban areas of Zahedan, 39, Journal of Geography and Development, No. 11.

General Meteorological Department of Mazandaran Province, 2010, thermal conditions in the cities of the province.

Azbon, Peter, 2001, Modernity in Transition from the Past to the Present, translated by Hossein Ali Nozari, second edition, Isfahan, Naghsh Jahan Publications

Akbari, Alir Gharkhloo, Mehdi, 2010, New Concept Ecotourism in Tourism Geography, First Edition, Select Publishing.

Amirani, Mohammad Hadi, 2003, Green Journey and Reflection on the Concepts and Benefits of Ecotourism, Jihad Monthly, Deputy for Extension and Exploitation System, No. 261, page 4.

Izadpanah Jahromi, Aida, 2002, Child of leisure and urban space, Master Thesis in Urban Planning, University of Tehran.

Ayafat, Amir, 1996, An Introduction to the New Concept of Environmental Tourism, Environmental Journal, No. 1.

Bahraini, Hossein, 1998, Urban Design Process, First Edition, University of Tehran Press.

Behroozfar, Fariborz, 2002, Introduction to Brakotourism, Payam Sabz Monthly, Second Year, No. 7.

Tavakoli, Habibollah, Ahadi, Farhad, 1992, Planning the use of Sorkheh Hesar lands for urban green space, Master Thesis in Urban Planning, University of Tehran.

Tolaei, Simin, 1373, The city and its environmental consequences, Quarterly Journal of Geographical Research, No. 33.

Haghani, Pouyan, 2002, Organizing the lands of Gisha Forest Park as a special area for tourism, Master Thesis in Architecture, Faculty of Fine Arts, University of Tehran.

Khademi, Mostafa, 1986, Green Space, Ministry of Program and Budget (Industrial Management), pp. 5-6.

Khaledi, Shahriyar, 1994, Application of Climate in Tourism Development of Iran, Collection of Tourism and Development Articles, Center for Tourism Research and Studies.

Rahimian, Kaveh, 1995, Ramsar Cultural and Recreational Center, Master Thesis, Tehran University of Science and Technology.

Zahedi, Shams al-Sadat, 2006, Fundamentals of Tourism and Sustainable Vacotourism (with Emphasis on Environment) First Edition, Allameh Tabatabaei University Press, page 4.

Zandpour, Mehdi, 2003, Histo-Ecological Tourism Center, Master Thesis, Tehran University of Science and Technology.

Management and Planning Organization of the country, 2001, Green space design criteria, (Office of Education and Development of Criteria) First Edition.

Cultural Heritage Organization of Handicrafts and Tourism, 2002, National Committee of Nature Tourism of Iran Compilation of national document on development and management of nature tourism in the country.

Handicrafts and Tourism Cultural Heritage Organization of Mazandaran Province, 2008.

Saeidnia, Ahmad, 1999, Municipal Green Book, Volume 1, Municipalities Organization Publications.

Sinaei, Vahid, 1996, Sustainable Development and Tourism, Monthly Political-Economic Information Monthly, No. 95.

Shaygan, Dariush, 2001, New Enchantment, translated by Fatemeh Valiani, second edition, Farzan Rooz Publishing.

Shahandeh, Behzad, 2001, Ecotourism is not only nature tourism, Green Wave Volume, No. 7.

Gilan Shahran Civil Company, 2008, Comprehensive plan of Shahran national tourism sample area, first volume.

Shariatnejad, Shamsollah, Sharifi, Morteza, 2009, Preliminary planning for ecotourism development, Quarterly Journal of Forests and Rangelands, Forests and Rangelands Organization, No. 34.

Shafiee Nasab, Seyed Reza, 2004, Design of Chitgar Lake recreational and water sports complex, Master Thesis, Imam Khomeini International University.

Shahraki, Ghodsieh, 2011, Location of tourist villages, Master Thesis in Geography and Tourism Planning, Islamic Azad University, Research Sciences Branch, Tehran.

Alambaz, Fatemeh, 2006, Chamkhaleh beach recreational tourist park, bachelor thesis, Tehran University of Science and Technology.

Kasmaei, Morteza, Ahmadinejad, Mohammad, 2003, Architectural Climate, Fifth Turn, Isfahan, Khak Publishing.

Kiarostami, Ghasem, The Necessity of Ecotourism Development to Prevent the Destruction of Valleys and Paradise Northeast of Tehran, Page 33, Payame Sabz Monthly, Third Year, No. 23.

Rahshahr International Group, 2009, Damavand Health Village Comprehensive Tourism Plan, Criteria, Spatial Regulations, Volume 2.

Gay Roche, 1991, Social Changes, Volume 1, translated by Mansour Vosough, Ney Publishing.

Majnounian, Henrik, 1995, Discussions about parks, green space and recreation areas, Parks and Green Space Organization of Tehran, first edition, Tehran.

Mohammadi, Maryam, Ostad Kalayeh, Forough, 2003, The role of traditional gardens in increasing the attraction of tourism and ecotourism, Payam Sabz monthly, third year, number 19.

Memarian, Gholamhossein, 2005, Introduction to Iranian residential architecture Extroverted typology, Tehran, University of Science and Technology Press.

Mansouri, Ali, 2002, Tourism and Sustainable Development, Journal of Geography Education Growth, No. 63.

Mirabzadeh, Parasto, 1996, Environmental Impact Assessment of Tourism Development, Mohit Rist Quarterly, No. 2.

Vashtasbi, Taghi, 2008, Architecture in Harmony with Climate, Amol, North Sustainable Publishing.

I want morebooks!

Buy your books fast and straightforward online - at one of world's fastest growing online book stores! Environmentally sound due to Print-on-Demand technologies.

Buy your books online at
www.morebooks.shop

Kaufen Sie Ihre Bücher schnell und unkompliziert online – auf einer der am schnellsten wachsenden Buchhandelsplattformen weltweit! Dank Print-On-Demand umwelt- und ressourcenschonend produziert.

Bücher schneller online kaufen
www.morebooks.shop

info@omniscriptum.com
www.omniscriptum.com